The First Freedoms and America's Culture of Innovation

The First Freedoms and America's Culture of Innovation

The Constitutional Foundations of the Aspirational Society

Narain D. Batra

ROWMAN & LITTLEFIELD PUBLISHERS, INC.
Lanham • Boulder • New York • Toronto • Plymouth, UK

Published by Rowman & Littlefield Publishers, Inc.
A wholly owned subsidiary of The Rowman & Littlefield Publishing Group, Inc.
4501 Forbes Boulevard, Suite 200, Lanham, Maryland 20706
www.rowman.com

10 Thornbury Road, Plymouth PL6 7PP, United Kingdom

British Library Cataloguing in Publication Information Available

Library of Congress Cataloging-in-Publication Data

Batra, N. D. (Narain Dass), 1937-
The first freedoms and America's culture of innovation : the constitutional foundations of the aspirational society / Narain D. Batra.
p. cm.
Includes bibliographical references and index.
ISBN 978-1-4422-2587-9 (cloth : alk. paper) -- ISBN 978-1-4422-2588-6 (electronic)
1. Freedom of expression--Economic aspects--United States. 2. Technology and law--United States. 3. Technological innovations--Law and legislation--Economic aspects--United States. 4. National characteristics, American. I. Title.
KF4770.B38 2014
303.48'30973--dc23

2013029639

The paper used in this publication meets the minimum requirements of American National Standard for Information Sciences Permanence of Paper for Printed Library Materials, ANSI/NISO Z39.48-1992.

Printed in the United States of America

In Memory of
My Mother
Whose Every Breath Was Love, Prayer, and Courage
THAKAR DEVI
1916–2012

Contents

Preface: How the Book Was Born

A casual, rambling, unscripted conversation struck with a friendly stranger at the Hanover Country Club one resplendent sunny afternoon in May a few years ago instigated me to think about how some cultures enable people to break the boundaries of the received tradition, or what Christopher Hitchens said in defense of Martin Amis, who was accused of racism in the wake of a London bombing of the public transport system, to "experiment in the limits of permissible thought," and keep their society intellectually alive, creative, and innovative. That golfing would lead to a Zenlike walking meditation and a discourse on creative permissiveness was an unexpected and stimulating experience. But that is the beauty of golf: sometimes it opens up doors that do not seem to exist.

Chung-Hee had just hit a mulligan and smiled in jovial embarrassment when I tried to look away. "Please join me though I am not very good," he said, as he extended his hand to introduce himself. With his dark glasses, muscular body, and vivaciousness, Chung-Hee looked like an older version of PSY (Park Jae-sang) of "Gangnam Style." We teed off straight and smooth, and at the fourth stroke we were both at the green. While I putted in two strokes, Chung-Hee stood patiently for his long putt, almost fifteen feet down slope. "I am going to do it, I am going to do it," he chanted softly to himself, as he looked at the ball and the hole from different angles, bending down and standing up with alacrity to read the green as if he were on a professional tour. The ball began to roll steadily and smoothly as he hit it with a relaxed, disciplined stroke, and it seemed to be going straight in but stopped for a split second and again began to slither and then again stopped at the lip as if to kiss the hole, but when Chung-Hee threw up his hands in sheer disappointment, the ball shook up and dropped into the hole with a swirling, clanking sound. Just like the beautiful girl I courted long, long ago in Seoul—no, no, no, but then finally she said yes. And we laughed so loudly that a squirrel that had scrambled up on the green scampered away. As we moved from stroke to stroke and hole to hole, wowing hits and regretting mishits, Chung-Hee's snatches of casual observations sounded like the tweets of an intellectual globetrotter. I was fascinated. The golf links had become like the Agora.

PROMETHEUS AND CONFUCIUS

"America is a country of unlimited desires," Chung-Hee spoke, expanding his chest. I asked, "Is that what brings you here?" "No and yes. Actually I am visiting my children, my daughter, a liberal arts maverick at Dartmouth, and son studying AI and Interactive Robotics at MIT. I have business interests all over—China, Singapore, India, and Boston, and I hope my children will return to Seoul, when they are done." "And you fear that they might not?" I asked clumsily. "Yes, I am afraid. America is so enchanting, so desirable, so seductive, like the Island of Circe. All I want my children to know is: why give up one country for another, South Korea for America, when you can have both? When I was growing up, and like other young Koreans, I was inclined toward Reverend Moon's Unification Church brand of Christianity, and my mother, who like most older-generation Koreans was a Buddhist, said to me one day: 'Why do you want to give up the Buddha for the Christ when you can have both?' That blew my mind. Trying something that is not permissible was very tempting, and my mother encouraged me to do so. So today I am a hybrid Christian-Buddhist-Korean, and I see everything interconnected and interdependent. America is so open that it creates limitless spaces for everyone, and so I feel at home both in South Korea and America, and I hope my children will do the same," he said, as we waited for another twosome in front of us to move on.

Nothing illuminates the Buddhist scientific-philosophical concept of interdependent coarising or dependent origination, Chung-Hee said, as the South Korean economic prosperity and democratic culture arising from its relationship with America since the Korean War ended in 1953. We have become so interwoven and interdependent, cooperating and competing, like Samsung and Apple, for example, that there is no end to opening up newer spaces, he said, newer possibilities being created for both countries. "But you also see the difference," I said, "don't you?" "Yes, indeed, Koreans, Japanese, Taiwanese, and Chinese are always trying to catch up by copying, cloning, and improving upon the models already created in the United States, and of course there's nothing wrong with that except that we have to wait and wonder as to what would be the next big thing," he said, with unusual candidness for an Asian man. "But why so, why can't you leapfrog, what makes America so attractive?" "Why, I don't know, perhaps because so many foreigners want to come to the United States to do things that they can't do back home, and that's what makes America so desirable."

Chung-Hee looked at me intensely and said, remember "the giant sucking sound"? Yes, I remember, the 1992 election, the presidential candidate Ross Perot railing against Bill Clinton about the North American Free Trade Agreement (NAFTA). "Few other countries," he said, "suck so much brainpower from all over world as the United States. Steve Jobs's Syrian father and American mother, Barack Obama's Kenyan father and American mother, for example . . . America is becoming more and more mixed up, mongrelized and hybridized. . . . more creative, more innovative, and more replenished every day, stronger and stronger. We Koreans love freedom," he said, "but we are afraid of free speech, extreme speech, fearless speech that allows Americans to think differently and break the boundaries of what is acceptable. We have Confucius but no Prometheus in Asian culture. We have the obedience and orderliness of Singapore but not the transgression and disorderly richness of Boston. I have been coming to America every year since I planted a business foothold here. I love to walk the streets of Boston. Every time I visit New York, I wonder how it would feel to be the mayor of New York. A Jewish mayor of New York! Unthinkable that this will ever occur in Beijing, Seoul, or Tokyo. If you happen to be in that corner of Asia, come and visit us in Seoul and see how different we are," he said, as we sat down for chilled, locally brewed beer at the clubhouse after a splendid four hours of golfing under the refulgent sun in the upper Connecticut River Valley. "Dartmouth College had just hired a Korean-American as its president," I said, and Chung-Hee glowed with pride.

HUMAN RIGHTS AND JULIAN ASSANGE

A few months later I got an opportunity to carry on and explore further my conversation with Chung-Hee about expanding the boundaries of permissible thought, fearless speech, and innovative behavior in the lecture I was invited to give at Dartmouth Master of Liberal Arts Studies, a program chaired by Donald E. Pease, a most distinguished international literary scholar and social critic. His recent book *Theodore Seuss Geisel* has been not only a pleasurable reading experience about an extraordinary creative genius of children's literature but also an exemplar of what Chung-Hee said of America's capacity and willingness to draw the brightest and the most creative people from all over the world. Dr. Seuss's grandparents were German immigrants, as were the space rocket scientists who came to America after World War II and enabled Americans to land on the moon, vide James A. Michener's fictionalized account and docudrama *Space*. Global brainpower and energy keeps flowing into America, thanks to its permissive culture and tolerance created by the First Freedoms. But the graduate audience at Dartmouth did not seem to be much enthused when I explained how the U.S. Supreme Court expanded the boundaries of permissible speech and thought through its gradual and progressive reading and interpretation of what the founding fathers intended to say while formulating the First Amendment. The question asked at the conclusion of my lecture was not about how free speech stimulates innovative behavior, the topic I labored to explore, but rather they seemed to be concerned about human rights, especially of the enemy combatants at Guantanamo Bay. Does a person lose his right to a fair trial under the U.S. Constitution once the government puts the label of enemy combatant on him? The question made me wonder as to whether a society can remain intellectually permissive and innovative if it forgets that human rights are not only inalienable but also must be universal.

We do ourselves great injury by denying human dignity and fundamental rights to anyone whether he is an enemy combatant or an illegal immigrant, and that's why fearless speech and uninhibited press, accountability, and transparency are indispensable to the civil society, the topic I explored as a panelist in a symposium on WikiLeaks held during the graduate residency program, an annual gathering of international students, at Norwich University, where I teach media law, criticism, ethics, and global diplomacy. With American media having become a business like any other big business seeking closer ties with the establishment, we need fearless antiestablishment whistleblowers such as WikiLeaks and Cryptome that cannot be intimidated or bought, I told the skeptical audience. The other panelists as well as the audience by and large were hostile to WikiLeaks, and I had to fight a lonely battle defending WikiLeaks as indispensable to not only keeping the government open and accountable but also for exposing the recurrent misdeeds of the untouchables of Wall Street and Big Global Finance. I said we need more nonstate, global whistleblower organizations like WikiLeaks regardless of whatever happens to Julian Assange, the founder of WikiLeaks who has been charged with sexual assault, and Bradley Manning, a U.S. Army soldier who has been charged with and has pleaded guilty to passing on classified documents to WikiLeaks. I enjoyed the confrontation with the young audience members, many of who had come from abroad, including the Middle East, to complete their graduate residency requirement at Norwich University. The freshness of the Arab Spring was still in the air, and someone asked if in the age of WikiLeaks and social media a country like Saudi Arabia, a staunch American ally, would survive. And someone in the audience yowled in protest when I said that no nation could survive for long when its women remained in bondage, an argument I had also made before a senior audience at the Institute for Lifelong Learning and carried on in my book, *Digital Freedom: How Much Can You Handle?* The graduate residency audience dispersed very much agitated as if someone had thrown a rock and disturbed a giant beehive. I thought I had done the day's work. After all, the job of a teacher, his moral obligation, is to disturb his students, not always to pacify them, so that they think on their own (Do I dare disturb the universe? In a minute there is time for decisions and revisions which a minute will reverse.—T. S. Eliot). As Chung-Hee said when we had finished putting at the eighteenth hole, the world needs more Prometheus than Confucius.

Dr. Seuss, Sergey Brin and Larry Page, Steve Jobs, Bill Gates, Francis S. Collins, Sam Walton, Jeff Bezos, Mark Zuckerberg, Julian Assange, and Oprah Winfrey are the children of the First Amendment, born of turbulence and harmony, challenge and response, culture and counterculture. No one could have summed up their élan, their passion, their spirit, their lust for life, better than the author of *Fahrenheit 451*, Ray Bradbury, who said, "The Universe says No to us. We in answer fire a broadside of flesh at it and cry Yes!"

Born amid the disciplined freedom of the golf course and the intellectual turmoil of the lecture halls, this is a book about the dynamics and dialectics of the aspirational society that creates constant innovation and breaks the pattern of history.

GRATITUDE

A good editor is a great blessing for a writer. Charles Harmon, a scholarly and perceptive acquisitions editor at Rowman & Littlefield, after retrieving the project from the lost-and-found bin, set out enthusiastically to see its speedy progress through various stages of internal and external editorial reviews and approval while at the same time keeping my head cool and spirits up. By wondering whether the title is good enough to embody the quintessential nature of the book, Mr. Harmon made me think again. Suddenly the old question, "What's in a name?" took a different meaning. The title of the book, especially its subtitle, *The Constitutional Foundations of the Aspirational Society*, came to me as a revelation that this is what I have been searching and discussing all along at various platforms, with my graduate and undergraduate students at Norwich University as well as my column readers at the *Statesman* and the *Times of India*. At Rowman & Littlefield, of course, there are many other capable editors who in various degrees have contributed to shape the book:

Chris Basso for his expertise in book production and managing the transformation of the manuscript to bound book including the aesthetics of the chapter and page design; Robert Hayunga for liaising with various departments and keeping the process flowing smoothly; Chloe Batch for the cover design that is creative, beautiful, as well as intelligent and sums up the book in a visual metaphor; and Desiree Reid for her knowledgeable and skillful copyediting of the manuscript, checking it for details, consistencies, and stylistic accuracies. But a beautiful book becomes meaningful only when it reaches the right audience for which I am thankful to Kim Lyons and her marketing team.

Unlike other fortunate professors in major research universities, I did not have the luxury of having research assistants to help me with the book, though I had a sabbatical that was very helpful. But having been a journalist given to working under the pressure of deadlines and in the habit of doing my own research, including interviewing and listening to people, I did as well as I could depending, however, upon online scholarly databases, and not least, Wikipedia, which I used judiciously as a launching pad to stimulate my thinking and pursue further research. Wikipedia is one of the greatest intellectual adventures ever taken, and this could happen only in America, where everything is in the making, in an unfinished stage, *Nullius in verba*, seeking disambiguation through collaboration and contribution, to reach a never-never stage of perfection.

Varsha, my wife of many wonderful years, a woman of extraordinary intuitive intelligence and strength of character, has been always there for me: the rock that I lean on and rest whenever I feel shook up. One could not have asked for a greater blessing in life. I completed the conclusion of the book in Chicago watching my bouncy little grandson Nayan ("the Lion," as his mother, Nikki, calls him) as he would run from one end of the apartment to the other, babbling, gesturing, pushing furniture, climbing up, upsetting and rearranging books and toys, while occasionally stopping to look at me with gusto.

The constant presence in my life of my son, Dr. Nik, with his quiet strength, dedication to work and ethics, unobtrusive manners, and enormous sympathy, respect, and concern for others, has been a great source of inspiration for me. And when I look at my grandson and my son, I see the future of America—free, innovative, and compassionate.

It is in this larger sense that *The First Freedoms and America's Culture of Innovation* is the manifestation of the principle of dependent origination, coarising from interdependent sources, geography, brainpower, and freedom.

Introduction: The Making of the Aspirational Society

The founding fathers, amid their epic struggle "to form a more perfect Union," one based on the self-evident truth that "all men are created equal; endowed by their creator with the right to life, liberty and the pursuit of happiness," sowed the seeds of the aspirational society of today. The aspirational society is an unending and relentless American experiment that creates the culture of innovation, one that is born of the extreme freedoms of the First Amendment. Consider this, for example: During the transformational changes that have been taking place in American society during the past few years, the First Amendment gave us the Tea Party's "We are the party of 'Hell, No,'" and Barack Obama's audacity of hope.

This kind of dynamic between turbulence and harmony, challenge and response, is the cradle, or perhaps we should say, the cauldron, of innovation, creativity, and self-renewal, which is the main focus of the book, and its most distinguishing feature. The triangular challenge-and-response system of the First Freedoms, open and competitive marketplace, and global brain-and-muscle inflow, creates spaces for Americans to become constant tinkerers, innovators, and entrepreneurs. Innovation is a kind of creative transgression, which breeds cascades of entrepreneurial opportunities that make the aspirational society.[1]

When confronted with a problem, an innovative mind conjures up something unique: new ideas, new inventions, and new programs based on unconventional approaches to solve the problem. But this can be done only if the political culture creates large breathing spaces by leaving the people alone, not as a matter of state policy but as something given, something inherent in being an American.

As a result, the First Amendment forbiddance against the state interference in the personal lives of the American people has encouraged fearless speech, unrestrained thought, iconoclasm, skepticism and irreverence for authority, endless experimentation, and continuous innovation, leading to newer developments in science, technology, the arts, and sociopolitical relations. Most of all, the First Amendment creates an environment of divergent thinking, nonconformity, and resistance to peer pressures. The First Amendment allows Americans to be outrageously creative, to be creatively crazy, to be impure, to be able to mix and remix ideas to create new technologies and new cultural forms and build new platforms to keep the great American experiment going.

We shall explore in this book critical factors that contribute to make the American society aspirational and innovative, and how it keeps renewing itself in the process. Societies in the past that were innovative, technologically and socially, were unable to sustain themselves because they had no self-correcting and self-renewing institutional system based on free and untrammeled thought. The Soviet Union, for example, achieved the notoriety of being the first to launch a satellite (*Sputnik*) into orbit and usher in the space era, and at the time of its dissolution in 1991, its space station was in full operation but its food shelves were empty.[2] As the allies discovered, Nazi Germany had space rocketry and modern electronics development in fairly advanced stages when the country was defeated, but its people were ill informed about themselves and the rest of the world.[3]

These societies lacked institutional systems of critical self-examination that enable a living social system to adapt and renew itself through the dynamics of challenge and response, the battle of ideas, and the internal checks and balances that free speech enables. Ideological dogma and dogmatic religion based on the claims of revealed truth, absolute truth, cannot stand the strict scrutiny of free speech. Free speech creates a tolerant society, a society that entertains all assertions of truthiness, as comedian Steve Colbert would call it, but lets such claims be tested in the cutthroat competition of the marketplace.[4] The 2012 presidential election, the reelection of Barack Obama, was truly speaking a mandate for acceptance. Acceptance is all—as the Bard would have said.[5]

The great American experiment began with the declaration that all human beings are born equal and free with certain incontrovertible rights, and that bold experiment has been continuing: through the Civil War culminating in the establishment of the Fourteenth Amendment; the women's suffrage movement; the civil rights movement; and the rise of an African American of mixed heritage and a Mormon businessman to the highest echelon of political life. And that great American experiment is still continuing in order to form a more perfect union of racially and culturally diverse people, the aspirational society, as we saw during the 2012 congressional electoral races.

And the great experiment of the aspirational society, like any other experiment by its very nature, must always be open to strict public scrutiny that brooks no hero-worship, no sacred cows, no revealed truths—the kind of rigorous investigation and accountability that can come only from an open, unimpeded, and robust press, especially needed in the age of intensely interconnected and codependent globalization.

> Those who make the mistake of thinking in terms of a first cause are fated never to become men of science. But as they do not know what science is, they are not aware that they are losing anything. To refer phenomena back to a first cause has ceased to be fashionable, at any rate in the West. . . . We shall never succeed in changing our age of iron into an age of gold until we give up our ambition to find a single cause for all our ills, and admit the existence of many causes acting simultaneously, of intricate correlations and reduplicated actions and reactions.
>
> —Aldous Huxley, *Ends and Means: An Inquiry into the Nature of Ideals and into the Methods Employed for Their Realization* (London: Chatto and Windus, 1941), 15–16.

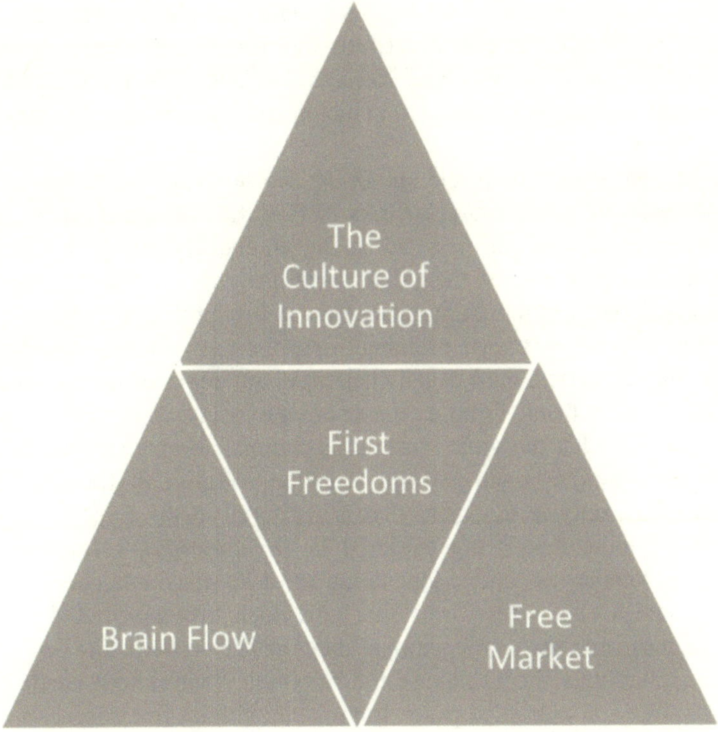

Dependent Origination of the Aspirational Society

During the past few years many important books on the First Amendment and innovation have appeared, and their authors talk about their fields of expertise with great insight and depth. None of the writers, however, discusses whether there is any correlation between the First Amendment freedoms and the culture of innovation, and the resultant emergence of the aspirational society. In this respect, *The First Freedoms and America's Culture of Innovation* will have a unique appeal for the reader at home and far from home.

Vijay Govindarajan and Chris Trimble, authors of *Reverse Innovation: Create Far from Home, Win Everywhere*, argue that innovation is no longer the exclusive domain of developed countries, and because of globalization and the rapid pace of economic development in other countries, locally adapted innovation at much lower cost is taking place everywhere.[6] They advise global corporations to develop their innovative products in developing countries and then bring them back to the United States and other developed countries, a process that they call *reverse innovation*, the equivalent of reverse engineering. The authors do not discuss the indispensable role of the critical culture provided by free speech, the free and fearless press, or the challenge of the marketplace in developing, testing, and evaluating innovation. Nor do they see innovation as a product of the aspirational culture arising from cascades of contiguous possibilities, endless spaces created by the First Freedoms. Nonetheless, *Reverse Innovation* is a brilliant how-to book for corporate CEOs interested in incremental improvement to stay competitive.

Clayton M. Christensen, the author of *The Innovator's Dilemma: When New Technologies Cause Great Firms to Fail*, has been a most influential thinker in the field of business and management, and the term *disruptive innovation* has now entered the global social parlance.[7] Initially a disruptive innovation appears in the marketplace with a low profile, almost stealthily, like a cult, offering new ways of doing things for a new market, but eventually it creeps into the old market and challenges and replaces the old technology of the existing firms. From a secular point of view, disruptive innovation is a process of what the great Austrian economist Joseph Schumpeter, perhaps inspired by the Hindu god Shiva, called "creative destruction," but it can take place only in a society that is anchored in the First Amendment freedoms, a topic that is not explored in *The Innovator's Dilemma*. The question is whether disruptive innovation can take place in a closed society, such as Saudi Arabia, where firms, like individuals, live a protected existence under a doctrinaire, theocratic national state security system. *The Innovator's Dilemma* overlooks the challenge-and-response environment created by a zealously free and unbridled press. Innovation without free speech is unsustainable.

Steven Johnson, in his book *Where Good Ideas Come From: The Natural History of Innovation*, discusses with his brilliant erudition and universal sweep as to what sparks trailblazing innovations from Charles Darwin's research of the ecosystem of the Galapagos that led to the Theory of Natural Selection to modern-day Silicon Valley innovations.[8] Johnson frames the innovation processes through several concepts, including the adjacent possible, fluid networks, the slow hunch, serendipity, failure-to-success process, repurposing an existing idea, and recombining and leveraging odds and ends into something new—important concepts that he explains with vivid examples. On the other hand, I argue in *The First Freedoms and America's Culture of Innovation* that the First Amendment provides Americans with the essential cultural environment, the primordial soup or the embryonic conditions, for innovations to develop regardless of the process, whether it is recombinant, serendipitous, or simply someone fiddling and tinkering in the garage or with the iPad.

The widespread innovative behavior at the grassroots level in America aggregates, congregates, amasses, and eventually spirals into some place called Silicon Valley that draws brainpower from all over the world. It is worth noting that two hundred years before the publication of *On the Origins of Species*, The Royal Society of London adopted its motto as *Nullius in verba* (Take nobody's word for it). All good ideas come from free and fearless speech, so go forth and discover. The world is an oyster: open it. (Heed the voice of Falstaff: I will not lend thee a penny. Pistol: Why, then, the world's mine oyster, Which I with sword will open.)

John Kao in his book *Innovation Nation: How America Is Losing Its Innovation Edge, Why It Matters, and What We Can Do to Get It Back*, like so many secular evangelists such as Thomas L. Friedman, Fareed Zakaria, et al., sounds an alarm that the United States, once the most innovative nation that decoded the genome, developed the Internet and Web 2.0, and gave rise to many great wonders of Silicon Valley, is now falling behind with dire consequences for the economy and national security.[9] Small countries such as Denmark, Finland, and Singapore are further ahead in innovation than the United States. Like other supernationalistic alarmists, Kao has a kind of implicit Marshall Plan: Mobilize national resources to put the United States ahead of the pack.

Innovation and creativity cannot be ordered, ordained, or sustained for long by any central plan or grand strategy, if the Soviet Union has left any lessons for us. Kao's book rightly emphasizes the need for investment in education and research, in science and technology, but it leaves the reader with the impression that innovation is the exclusive domain of the elites. Kao does not highlight the role of culture in nurturing long-term innovative behavior generated by the First Freedoms that have created the aspirational society from the bottom up. In passing, one might wonder as to how Singapore compares with New York City in creating the culture of innovation. A few years ago when I visited Singapore, I could hardly stay and breathe in the clinically pure air of the city-state free from free speech and other intellectual pollutants. It was such a relief when I returned to the noise and chaos of the streets of New York City.[10]

Historian Niall Ferguson, the author of *Civilization: The West and the Rest*, writes with a wide angle and gives us a vast and intriguing sweep of history as to how the West surpassed the glory of the East and the Arab-Muslim world by developing six "killer applications," including competition, science, democracy, medicine, consumerism, and the work ethic.[11] The West has no monopoly over these applications and therefore any country that can download these killer applications could compete and surpass the United States and Western Europe, he argues, as China is doing now. The most difficult "killer application" to develop is democracy, a system of decentralized and divided powers based on the rule of law and the free press, which cannot be downloaded like a computer program, developed in haste, pirated, or reverse engineered. As I argue in *The First Freedoms and America's Culture of Innovation: The Constitutional Foundations of the Aspirational Society*, the First Amendment freedoms are the most difficult ideas to grasp, practice, and sustain, and therefore the rest of the world, including China, cannot catch up with the West unless they incorporate the First Freedoms as the ruling doctrine.

The First Freedoms and America's Culture of Innovation challenges the proposition that America's decline and fall is inevitable. Given the necessary and sufficient conditions, the aspirational society need not decline and fall. When you have millions and millions of people at the grassroots level habitually tinkering and innovating and expanding their sense of freedom, it is difficult to imagine the society's decline and fall, though there may be periodic ups and downs, sometimes extremely painful man-made, near-death-like conditions—for example, the Great Depression, the Dust Bowl, or the financial collapse of 2008. But every catastrophe has made America stronger, smarter, and more humane, thanks to the First Freedoms. If history has a pattern of decline and fall, then the First Amendment is the monkey wrench that breaks the pattern.

PROMETHEAN POSSIBILITIES OF THE ASPIRATIONAL SOCIETY

Free speech, political humor, parody, and satire keep American society from tripping over to extremism. The strength of America, its capacity for self-renewal and innovation, comes from its secular scriptures, the First Freedoms, which have been expanding since the U.S. Supreme Court began to examine them very early in the twentieth century. The function of free speech is to free the American people from irrational fears, opined Justice Louis Brandeis in a landmark case. The First Amendment frees the mind to search and seek answers wherever it can. It liberates the mind to entertain the idea of transgression. Creative transgression more than creative destruction engenders innovative behavior. Freedom from irrational fears has enabled Americans to develop into a nation of tinkerers, innovators, and experimenters. Nevertheless, to rectify the excesses of the aspirational society, the First Amendment has created a system of self-correction through a continuous stream of critical social discourse in the media, which by and large have been bold and unconstrained.

The dialogue about innovation in social and biological fields continues unabated in the United States, thanks to free speech that makes the exploration and the articulation of the profane and the sacred such as embryonic stem cell research and surrogate motherhood any which way possible. For women, for example, freedom from the chains of menopause and reproductive equality with men are two major gains from new biotechnological developments, which would not have been possible without freedom of (and from) religion, freedom from revealed truths and religious taboos, as guaranteed by the First Amendment.

The American innovator is the product of a cultural environment that has been evolving for more than two centuries and is predicated upon minimal government control and interference. Although the First Amendment commands governmental noninterference into the personal lives of the people, it does not direct the government for creating infrastructure to augment innovation culture except through copyright and intellectual property laws. The government has no doubt played an indispensable role in the development of the Internet, for example, but its commercialization, democratization, and globalization were left to tinkerers, innovators, and entrepreneurs.

The research of European scholar Mario Coccia of the National Research Council (Italy) suggests a strong correlation between democratization and technical innovation. According to his research findings, liberal, participatory, constitutional democracies are more creative than autocratic societies due to the fact that openness and the rule of law of free countries encourages a high level of interaction among social and economic factors that eventually lead to invention, economic growth, and wealth creation, an example of sustainable, interdependent coarising of innovation.[12] The openness of the United States puts it in an incomplete mode, in search of a more perfect union, the aspiration society always in the making. The perfect union, the final product, has yet to come if the flow of information and the news media remain unencumbered, vigorous, and wide open.

A vigorous and wide-open newsosphere is a necessary antidote to the excesses of the aspirational society. The business of chronicling the world honestly, objectively, and accurately is a serious business, and we need both professional as well as amateur journalists to trawl the globe and make it knowable, transparent, and governable, consistent with the free will of the people. The question is: Will the press continue to be "unfettered, robust, and wide open" for democratic deliberations, keeping a check on unbridled corporate and political power concentrated in the hands of a few, and for spurring self-renewal and innovation?

The future of the aspirational society depends upon how vigilant are the new media since the traditional media have the temptation to become an invisible part of the establishment. As Lee C. Bollinger says, "Globalization intensifies our need for the press to remain free and independent so that it can report accurately on the world, from the world, to the world. . . . How will we confront this new, more reactive, and more integrated world with a press weakened by a lack of stable funding and forced to navigate through a bewildering landscape in which the laws governing censorship and access to newsworthy information vary from nation to nation?"[13]

The challenge to the authoritative voice of the traditional news media and their claim to infallibility, their prerogative of serving the public interest through investigative reporting, opinion-making and agenda-setting functions, comes from various sources including technologically empowered independent citizen journalists and bloggers, news aggregators and opinion websites, as well as emerging new online, nonprofit organizations such as WikiLeaks and Cryptome that provide bulletproof anonymity to whistleblowers and leakers of unnecessarily classified documents that hide incompetency and corruption. It is the challenge of the ever-expanding environment of news ecology that we have to keep in mind while evaluating how indispensable are the old traditional media for the free flow of information, openness and transparency, democratic self-renewal, and the culture of innovation in the aspirational society.

Commercial speech, although regulated to protect the consumer, is a most notable hallmark of the American expressive behavior. Free speech and free flow of information imbue the marketplace with dynamic energy that propels business enterprises to compete for customers through scintillating commercial messages in various media. Advertisement stimulates needs, wants, and desires; it expands the marketplace, and it encourages businesses to innovate for competitive advantage. By arousing wants and desires, by transforming wants and desires into compelling needs, commercial speech is one of the greatest sources of disruptive innovation in the aspirational society.

Commercial speech is not only the foundation of American consumer culture but it also instigates social changes and stimulates innovation and creativity in myriad ways. The soft power of American aspirational brands has been a noticeable source of social and political changes in many countries. On the other hand, excessive patenting in the burgeoning field of e-commerce could, however necessary, diminish the innovative power of commercial speech. Similarly, excessive expenditure on corporate speech to influence political decisions has not only diminishing returns, as we saw in the 2012 elections, but also diminishes aspirations for innovation.

Marketplace fundamentalism is a noticeable manifestation of the aspirational society. The laissez-faire extreme form of capitalism implies that the market forces of supply and demand can solve all problems, and its believers credit it with fuelling unprecedented economic growth. Unfortunately, unchecked and unregulated marketplace extremism has bogged down corporate America into financial scandals, triggering a loss of faith in the marketplace. When corporations grow big, fat, and rich, it becomes difficult for them to keep their surplus capital secure from the greed of corporate executives and politicians. Some of it illicitly flows into the pockets of the people who control them. These people become victims of their successes and excesses, and finally meet their hubris.

Government rules and regulations alone cannot correct the excesses of the marketplace. The news media have indeed been playing an important role in exposing corporate corruption in the financial industry and criminal negligence in the pharmaceutical industry, for example, but the news media do so after the damage has been done. Social media, including WikiLeaks, are crucial in keeping a constant watch in real time on the marketplace and checking corporate behavior.

To protect themselves from public calumny, corporate boards should establish the equivalent of the Homeland Security apparatus to preempt corruption at the aspirational stage rather than discover corruption when it has already happened. Moreover, since the traditional news media organizations are either incapable of understanding the convoluted workings of Wall Street firms and complexities of global corporations or are prone to collusion with them through bribery, inside trading, careerism, or intimidation, the American public should not hesitate to welcome and embrace a free-floating, footloose, supranational organization such as WikiLeaks. Corporate frauds, to a great extent, can be checkmated by media vigilance. But since corporate America uses China as a manufacturing hub, the American news media must not leave China alone. Corruption in China affects corporate America and investors.

China's market-oriented authoritarianism challenges American democratic exceptionalism. The question is whether a half-free society can ever challenge American exceptional powers that grow out of the First Freedoms. Freedom of the press will become a necessary condition for future market-based economic growth in China. Censorship is an economic bottleneck because it restricts the flow of valuable information and warps the information present in the marketplace, thus hampering the growth of markets, particularly financial markets, according to Zhiwu Chen of the Yale School of Management.[14]

However, in the age of the Internet with its decentralized freedom, many-to-many spontaneous and instantaneous twittering and social networking, free-wheeling bloggers, and the medium's drumbeating redundancy, the challenge for the Chinese authorities is how to use new media for collective mind control lest millions of dispersed voices converge into flash mobs to oppose the government or solidify into a political opposition. But commercialized media seeking audiences and advertising support from the open market might become subversive in thousands of small and unobtrusive ways. The Central Communist Party might lose public backing. This new dynamics between the one-party authoritarian government and market-oriented media struggling for greater freedom, especially when authorities want to use the news media for gauging public opinion and surveillance functions, will have tremendous consequences for China. What has kept the Chinese political and economic model viable so far is the vested interest of global corporations—especially corporate America—to use China's docile and cheap but hardworking workforce for contract manufacturing. Authoritarian China cannot handle political freedom. China does not have American cultural capabilities, its open spaces, without which it cannot establish its global supremacy and challenge America even if it continues growing at the present rate. China cannot control the integrated global economy of which it is an integral part—no one can.

To live and die in freedom, that is all about America, but what about the rest of the world? In the Second World War Americans (and their allies) crushed Japan and Germany. Subsequently, they turned them into civilized trading partners, creating a sphere of global coprosperity. Germany and Japan will never be a military threat again even if they keep growing economically. And that was the unyielding resolve after 9/11—cold rage against terrorism, the international menace that must be eradicated. The Department of Homeland Security and the PATRIOT Act, however controversial, became innovative tools of securing the homeland as Americans conducted war and diplomacy. Quick successes led to neoimperialistic views in some conservative political and academic quarters. The neoimperialist vision of the world and the political situation is that a most dominant political power, the sole superpower, the United States of America, whose GDP is more than the total GDP of the next two economic powers combined, must establish benevolent hegemony and do nation building in order to maintain world peace.

But to build a modern civic society, it is necessary to erect pillars of freedom: freedom of expression, freedom of religion, and most importantly, equal opportunity for women. Without women being free and equal in all respects, half-legged freedom cannot survive for long. Nor can family life be productive without women being treated equally at home and in the workplace. Without the essential institutional bulwarks of human freedom, a country may become economically prosperous, may have open markets and free trade, but still could have closed minds fit for tyrants to rule.

The magnitude of the problem that the United States has been facing for nation building abroad is tremendous. It amounts to bringing about a cultural shift. But the United States, too, has undergone a massive cultural shift as it has moved from George Walker Bush to Barack Hussein Obama, a continuous manifestation of the dependent origination or the interdependent coarising of the aspirational society, which is, as President Obama said at the second inaugural, "Our celebration of initiative and enterprise; our insistence on hard work and personal responsibility, are constants in our character. America's possibilities are limitless, for we possess all the qualities that this world without boundaries demands: youth and drive; diversity and openness; an endless capacity for risk and a gift for reinvention."

In the rest of the book, we shall explore America's Promethean possibilities created by the culture of innovation that emerges from the incessant interactions among the First Freedoms, the free market, and brain inflow.

Chapter One

Turbulence and Harmony

THE CAULDRON OF CREATIVITY

The aspirational society is an unending American experiment that creates the culture of innovation by breaking down psychological barriers, social inhibitions, and freeing people from irrational fears. Sacred taboos, conformist thinking, ideological dogmas, and establishmentarianism create irrational fears. Failure in the aspirational society does not create a corrosive sense of shame because there is always another chance.

First Amendment scholars and jurists have been propounding the view that the most important rationale and the primary goal of free speech is civic participation, democratic governance, and testing truth in the marketplace of ideas. But free speech has another important function without which a democratic society will stagnate and become a tower of babble. It is the function of free speech to liberate people's imagination and dare them to do things that no one has done before, go "boldly where no man has gone before," which is the culture of the aspirational society.[1] Or face the consequences.

The consequences of not facing the challenge of *Sputnik* in 1957, for example, would have been terrible for the United States even though it would have continued to remain a democratic country. There would have been no America, the aspirational society, as the world looks up to and wants to emulate it today. "Those who came before us made certain that this country rode the first waves of the industrial revolutions, the first waves of modern invention, and the first wave of nuclear power, and this generation does not intend to founder in the backwash of the coming age of space," John F. Kennedy said.

> We mean to be a part of it—we mean to lead it. For the eyes of the world now look into space, to the moon and to the planets beyond, and we have vowed

> that we shall not see it governed by a hostile flag of conquest, but by a banner of freedom and peace. We have vowed that we shall not see space filled with weapons of mass destruction, but with instruments of knowledge and understanding. . . . We choose to go to the moon. We choose to go to the moon in this decade and do the other things, not because they are easy, but because they are hard, because that goal will serve to organize and measure the best of our energies and skills, because that challenge is one that we are willing to accept, one we are unwilling to postpone, and one which we intend to win, and the others, too.[2]

John F. Kennedy, like other great American leaders, conceived American democracy as the aspirational society, one based on the dynamics of challenge and response.

By its very nature the aspirational society is based on ever expanding and unexplored spaces that are continuously being generated by First Amendment freedoms. There is no generative algorithm for the aspirational society except that the First Amendment fuels freethinking behavior and free market entrepreneurship, which produce not only social flexibility but also enable the global flow of brain and muscle power into the United States. The Kennedys, who were driven from the hardscrabble and famine-stricken Ireland to become a most famous political dynasty in America, and the Mormons, who were driven from place to place to create a new theological and spiritual space for themselves, for example, had the same aspirations as the steady streams of immigrants who are huddling the borders and shores of America today. In the imagination of an immigrant, America is the country of the First Freedoms, where an Austrian immigrant, such as Arnold Schwarzenegger, whose family had an ambiguous Nazi past, for example, can go to the top, become rich, famous, and powerful, and at the same, perhaps, wonder why he would end up as a pathetic figure seeking redemption for his excesses. Schwarzenegger did not learn any lesson from the excesses of the magnificent Kennedys, the family into which he married. Freedom in the aspirational society demands accountability. There is no forgiveness, though there may be a second chance.

Being aspirational is a universal human condition and implies imagination and initiative, and it requires venturesome, individualistic, and aggressive behavior—the kind of human condition that cannot be unleashed by the state. Rather, the state should leave people alone to their own devices as the founding fathers wrote it into the First Amendment.

Like the informational molecules that contain genetic instructions for the growth, development, and functioning of a living organism, each society has its DNA. The aspirational and experimental nature of the American society is coded into its DNA, which is the First Amendment. One might wonder what kind of DNA the Chinese society has, the informational code that shapes it; it is certainly not the First Freedoms. But keeping in mind the evolutionary biology and its broad application to the evolutionary nature of human society, a society's DNA can slowly change in the course of time. Every society has the potential to gradually assimilate the First Freedoms into its system of core values and become the aspirational society, open and innovative, where possibilities are limited by one's imagination.

Interestingly enough, there are no special aspirational classes in America because aspiration permeates through every echelon of the society. People at the bottom of the pyramid, including newly arrived immigrants, legal and illegal, have as much aspiration to achieve and excel as people on Wall Street. And they know that there are many avenues to reach there, whatever space they imagine, whether sports, politics, business, managerial professions, Silicon Valley, or other unexplored imaginaries yet to be named. It is the First Amendment that makes it possible to explore "adjacent possibilities," the unopened doors that did not seem to exist but have been there, endlessly expanding as one continues exploring. The aspirational society expresses a holistic view of life that if a person is not blinded by doctrinal and dogmatic thinking, whether religious or ideological, it is possible to create spaces that never existed before, which is at the heart of the American experiment, the journey toward a more perfect union. The aspirational society is not a race to occupy the top space for the simple reason that space at the top can be created and expanded, and therefore there are no losers in the race to the top. It is not a zero-sum game because by creating opportunities, one might say, at the risk of mixing metaphors, that in the aspirational society, all boats rise.

Pragmatism

What Jon Meacham said about Thomas Jefferson is true of most Americans, most of the time. Meacham said that Thomas Jefferson "was not at all handcuffed by ideology; if he believed it would serve the American cause, he would do just about anything."[3] From the Louisiana Purchase in 1803 through Nixon going to China in 1972—the two metaphors most succinctly illuminate what America is all about—Americans have always used pragmatism as a tool of innovative behavior. Thomas Jefferson expanded America to the Pacific Coast just as Richard Nixon expanded America to the China Sea. The aspirational society is a pragmatic society without boundaries. Walmart, Microsoft Windows, Google, Facebook, or Twitter have no boundaries because they are the product of the same bold imagination born of the First Freedoms as the Louisiana Purchase or Nixon appropriating communist China into capitalism.

The people of the aspirational society are the First Amendment evangelists who speak through their actions, failures, and successes. Free speech is their creed and the tool to explore, to break, to rebuild, and to renew. Considering the history of free speech from the times of the ancient Greeks, the long European Dark Ages, the invention of the Gutenberg printing press, the rise of Martin Luther and Protestantism, the rediscovery and the reappropriation of the Greco-Roman rational spirit of inquiry during the Age of the Enlightenment, and the American continuous, everlasting struggle with the First Amendment, it is difficult to conceive free speech as a mere app that can be easily downloaded or pirated by any nation. Niall Ferguson needs to think again as to why the West rose and why it is not going down.[4] But at the same time, keeping in mind the long interregnum of the Dark Ages, every society has the potential to fall into the heart of darkness. Freedom is not an entitlement. It must be the core value that informs the aspirational society.

Blind faith in religious dogma does not serve the American national interest or advance the American experiment toward a more perfect union, and that is why in the 2012 presidential election American voters ignored the Billy Graham Evangelistic Association's desperate plea to voters in a full-page ad in major newspapers a few days before the election, in which the Reverend Graham said, "On Nov. 6, the day before my 94th birthday, our nation will hold one of the most critical elections in my lifetime. We are at a crossroads and there are profound moral issues at stake. I strongly urge you to vote for candidates who support the biblical definition of marriage between a man and a woman, protect the sanctity of life and defend our religious freedoms. The Bible speaks clearly on these crucial issues. Please join me in praying for America, that we will turn our hearts back toward God."[5]

It must be a terrible disappointment for the Reverend Graham to see the America of his religious imagination disappearing before his very eyes. The aspirational society is experimenting with a new idea of marriage, an idea that—albeit fully protected under the Equal Protection of the Fourteenth Amendment—has yet to stand the test of time and prove as much functional as the traditional marriage. Americans are a down-to-earth, no-nonsense, hardheaded, practical people. If the idea of nonbiblical marriage works, the American people will accept it regardless of what the Bible or the Supreme Court says. Like the late Reverend Jerry Falwell, the Reverend Billy Graham, too, has failed to accept that at the heart of American society is the First Amendment, not the Bible. Most Americans cherish the Bible, though very few read it. Americans are, broadly speaking, a very religious people, but at the same time they are fond of experimenting, blending, and mashing up different religious and secular traditions—yoga and Christmas, Thanksgiving and Black Friday—thus increasing tolerance and acceptance.[6]

But orthodox religion, the crusading spirit, the religious militancy, does not run deep into their veins, though one might see its occasional manifestation here and there. Religion does not determine their daily decisions. American behavior, unselfconsciously, is propelled by the First Freedoms. Few other peoples have the capacity for tolerance for what psychologists call *cognitive dissonance*, the ability to live with contradictions, as Americans do, thanks to the First Freedoms. The First Amendment is the enemy of sacred cows. Wall Street, the military, the judiciary, the political establishment, or the church cannot prop up false heroes for too long because the news media are unforgiving, and the American people are not given to hero-worship, though they admire heroic virtues, as they did in the case of General David Petraeus, until it was found that the general's personal conduct did not match his military valor. General Petraeus, the son of a Dutch immigrant father and American mother, is a product of the aspirational society, and he could have a greater future.[7] General Petraeus did his job, as other generals have done in the past, and it is time to forget him. There will be other heroes to be held accountable. America is an iconoclastic society. Without accountability, the aspirational society will become an acquisitive, rapacious society.

PURSUING THE FREEDOM TRAIL WHEREVER IT LEADS

"Everyone shall give an account of himself to God." That's what a giant billboard admonishes visitors as they drive through the verdure vastness of the Pennsylvania Dutch countryside.[8] Unabashedly the U.S. dollar bill says, "In God We Trust." Every oath-taking ceremony ends with "So help me God!" and "Amen!" repeated after the person who administers the oath. Sneeze, and someone will say, "Bless you," even if the person is a stranger. It cannot go unnoticed that the American people have been observing the anniversary of the September 11 terrorists' attacks under a patriotic trinity: God, the USA, and Terrorism. The trinity, the psychological equivalent of space-age global positioning systems, enables the American people to be aware of the threat to their existence, the American Exceptional Way, relative to the known positions and undeclared intentions of America's ubiquitous enemies, a positioning system that makes Americans see the world uniquely, whether they support a Democrat or Republican candidate for Congress or the White House.

A first-time visitor, expecting the United States to be a continental-size Singapore plus Hollywood and Las Vegas, might feel that there's too much of God here, but the first cultural shock soon wears away because the American people don't give much deeper significance to it. Perhaps Americans invoke God often as a mode of phatic communication, or because sometimes they have nothing else to say. Most Americans call upon God as a social crutch, much as a Frenchman would say *Bonjour* (which means, as perhaps comedian Steve Colbert would say, that French wine and women are the best!), unless Americans feel stressed and threatened, as happened in the aftermath of the September 11 terrorist attacks when they mixed God with patriotism to make such a heady booze that it blew everyone's head off, and it turned the nation into one closed, collective mind listening to the drumbeat of apocalyptic fear emanating from the White House. It is also true, more than ever before, that American political discourse, especially during the campaign season, as we saw in the Republican presidential primaries, is suffused with references to God despite the constitutional brick wall between the church and the state. You can't live in fear and be free, and so invoking God, perhaps, makes you feel free from fear. But to be free in America means making your choice, which includes choosing your own God—monotheistic, polytheistic, or nontheistic—or even the God-particle, Higgs boson, that the CERN lab in Switzerland has claimed to discover.

Court Ruling

It is not the government's business to tell Americans what God to choose, or not to choose at all, thus speaks the First Amendment, and so ruled the U.S. Court of Appeals for the Ninth Circuit on June 26, 2002. The decision came after a lawsuit filed by a physician, Dr. Michael Newdow, an atheist who complained that his elementary school daughter's First Amendment right to be free from God were violated when she was given no choice but to "watch and listen as her state-employed teacher in her state-run school leads her classmates in a ritual proclaiming that there is a God, and that ours is 'one nation under God.'"[9] That does not mean that atheists have no place in these United States, except that they should not take their atheism and church-state separation wall too seriously.

Although not all states require the Pledge of Allegiance to be included in the school day, half of the state legislatures have made its inclusion obligatory. Making the situation foggier is a 1943 U.S. Supreme Court decision that children cannot be forced to recite the pledge, though the school's public address system may recite it.[10] In matters of religion, the government shall remain uninvolved; that is the constitutional mandate. Until 1954, the Pledge of Allegiance included the phrase *one nation indivisible*, but to fight the godless communism, the Soviet or Chinese variety, Congress changed the pledge to include *one nation under God.*[11] The wall between church and state, argued some extreme secularists, has cracked a little bit. And since then God, not a sectarian one, Catholic or Protestant, but a generic God, a mythical Supreme Deity, the God of the Declaration of Independence document, has found a frequent place in public discourse. In a two-to-one ruling, Judge Alfred T. Goodwin of the Circuit Appeal Court ruled that the pledge that we are a "nation under God" is identical to saying that we are a nation "under Jesus," a nation "under Vishnu," a nation "under Zeus" or a nation under Zarathustra.[12] Such a profession violated the First Amendment mandate that the government shall stay neutral in matters of the establishment of religion or the exercise thereof. It would send a wrong message to impressionable children that if they did not participate in the recitation of the pledge, they might become "outsiders," the judge wrote.

The court discovered a glaring contradiction in what the U.S. Constitution established and what the American people want to believe and practice. The pledge excludes polytheists who worship multiple gods and goddesses; atheists and agnostics; or those like the Buddhists, who believe in nirvana, a state of supreme bliss, tranquility, and purity that is attained when the self is absorbed into the Infinite. It has been rightfully argued that the recitation of the pledge that includes "one nation under God" is psychologically coercive, because it insidiously shoves people to accept monotheism as the sole religious path available to them. The pledge not only violates the First Amendment but also negates cultural pluralism. It violates freedom of choice, the essence of being an American. Americans should return to the original wording in the pledge, "one nation indivisible," so it is argued, because it is inclusive as well inoffensive. However, most Americans, like the dissenting Judge Ferdinand F. Fernandez, dismiss the fear that the United States might become slowly and gradually a theocratic state because of the inclusion of *under God* in the pledge. The appeal court's ruling was overturned by the U.S. Supreme Court decision in 2004. Justice William Rehnquist, writing for the majority, said, "To give the parent of such a child a sort of 'heckler's veto' over a patriotic ceremony willingly participated in by other students, simply because the Pledge of Allegiance contains the descriptive phrase 'under God,' is an unwarranted extension of the establishment clause, an extension which would have the unfortunate effect of prohibiting a commendable patriotic observance."[13] Thus God, though not of the Bible, has become part of American patriotism.

BUT THANK GOD: FOR EVERY JERRY FALWELL THERE'S LARRY FLYNT

Consider the global power of the American news media. A Hindu-Muslim flare-up killed eight innocent people thousands of miles away in Sholapur in western India in 2002, when a Southern Baptist preacher, goaded by the CBS *60 Minutes* confrontational-style ambush interview, ridiculed Islam.[14] Odd it might seem, nonetheless this is a despicable facet of U.S. media serving the globalized marketplace unprepared for the extreme form of free speech, which Americans take for granted. CBS correspondent Bob Simon should have known the self-righteous attitude of the late Reverend Jerry Falwell and anticipated the kind of bigoted response he would get when he asked Falwell the loaded question about the Prophet Mohammed. To aggravate the matter and compound the insult, CBS went on drumbeating Falwell's disparaging remark to multiply the audience for the Sunday night prime-time broadcast, which naturally did not go overlooked by Muslims all over the world. In this era of intensified and ragged religious sensibilities, when American leaders can't stop assuring the Muslim world that the fight against terrorism is not a clash of civilizations, CBS could have chosen to balance its hunger for market ratings in the name of First Amendment freedoms with the ethical responsibility by inviting a Muslim cleric to respond to Falwell. But reporter Simon let hatred spill from the mouth of the man who had a record of vitriolic, intolerant, and bigoted attitudes against anyone who did not agree with his extreme right-wing, Christian, dogmatic views.

Soon after the September 11 terrorists' attacks, Falwell went on a Christian broadcast network, *The 700 Club*, and uttered the most reprehensible words unbecoming a preacher: "I really believe that the pagans, and the abortionists and the feminists, and the gays and the lesbians who are actively trying to make that an alternative style, the ACLU, People for the American Way, all of them who have tried to secularize America, I point the finger in their face and say, 'You helped this happen.'"[15] In 1999, he denounced a widely watched British children's television show *Teletubbies* because he believed Tinky-Winky to be homosexual.[16] Falwell wasn't the only American preacher to have used broadcasting media to fuel hatred against Islam. There have been others, too. Several theologians and preachers, for example, the Reverend Pat Robertson and the Reverend Franklin Graham, son of evangelist Reverend Billy Graham, of different Christian denominations and with global television ministries, have used the pulpit to disparage Islam and the Prophet, which provoked the Iranian leaders to issue a clarion call for the killing of the three ministers. But killing is no way to fight intolerance in the United States. Americans know better how to keep religious and political leaders in their places. In America humor fights bigotry and keeps the country levelheaded.

But there is no better way of dealing with America's bigoted holy men than through untrammeled free speech, ridicule, parody, and satire, the way America's most celebrated pornographer Larry Flynt of *Hustler* magazine trashed Falwell when the evangelist headed what was called the Moral Majority of the Ronald Reagan era. In 1983, *Hustler* published a parody of an advertisement for Campari liquor, titled "Jerry Falwell talks about his first time." Mimicking the format of Campari ads, *Hustler* carried a mock interview with Falwell, who recounted his "first time" as a drunken sexual encounter with his mother in an outhouse. Although at the end there was a disclaimer ("Ad parody—not to be taken seriously"), this was a most shocking and repulsive satire about a religious bigot masquerading in the attire of an evangelist preacher. (Falwell: "Oh, yeah. I always get sloshed before I go out to the pulpit. You don't think I could lay down all that bullshit sober, do you?"[17])

When the U.S. Supreme Court took up the case in 1987 as to what to do when free speech causes invasion of privacy and inflicts emotional distress on a public figure, Chief Justice William Rehnquist, writing for the unanimous Court, acknowledged that the parody was "gross and repugnant," nonetheless, he said, parody is the privileged domain of the satirist and cartoonist and is protected under the First Amendment.[18] *Hustler*'s portrayal of Falwell as an incestuous drunk was so outrageous that few would believe it, he said on behalf of the Court:

> This case presents us with a novel question involving First Amendment limitations upon a State's authority to protect its citizens from the intentional infliction of emotional distress. We must decide whether a public figure may recover damages for emotional harm caused by the publication of an ad parody offensive to him, and doubtless gross and repugnant in the eyes of most. Respondent would have us find that a State's interest in protecting public figures from emotional distress is sufficient to deny First Amendment protection to speech that is patently offensive and is intended to inflict emotional injury, even when that speech could not reasonably have been interpreted as stating actual facts about the public figure involved. This we decline to do. At the heart of the First Amendment is the recognition of the fundamental importance of the free flow of ideas and opinions on matters of public interest and concern. . . . As Justice Holmes wrote, "when men have realized that time has upset many fighting faiths, they may come to believe even more than they believe the very foundations of their own conduct that the ultimate good desired is better reached by free trade in ideas—that the best test of truth is the power of the thought to get itself accepted in the competition of the market."[19]

This is the way American society anchored in the First Freedoms deals with public monstrosities such as Falwell. It is through the ridicule, laughter, parody, and satire of late-night shows, stand-up comedians, and smutty magazines that American society corrects and controls the pompous, the self-important, the arrogant and the pretentious, both on the extreme right and left, and thereby maintains its sanity and dynamic equilibrium. For every bigoted holy man in the United States, there is a gentleman-pornographer who shouts that God is Sex, Sex is God, and folks, "That's all ye need to know."[20] Extremes in American society repudiate extremes, keeping the nation sane, secular, free, and open, as the Founding Fathers imagined that it ought to be. Only a tolerant society can accommodate the Jerry Falwells and the Larry Flynts and maintain its dynamic equilibrium and remain innovative and creative.[21]

CRUDE AND OUTRAGEOUS SPEECH AS CORRECTIVE

In this hypersexed society, of *Vagina Monologues* and *Sex and the City*, "Girlie Men" and "Bushwhooping" women, the excitement of verbal sparklers is part of the social discourse.[22] Consider this, for example. During the 2004 election campaign, a John Kerry Democratic supporter from Hollywood compared George W. Bush with her private parts, and another one from the state of "Kaalifornia," an immigrant muscle-builder-actor-turned Republican politician, dubbed his political opponents as sissies. Perhaps befuddled Alice in Wonderland would have described the American domestic political scene during election time turning rather "curiouser and curiouser." When Arnold Schwarzenegger, a man used to getting his way with women every which way, became governor of California with high hopes but failed to persuade the Democrat-dominated state legislature to approve his $103 billion proposal for the fiscal year, he lost his self-possession as well as civility and screamed at his opponents, the "girlie men," and admonished Californians to "terminate" them in the coming November 2004 elections.

The Austrian-born American dreamer and a rich Hollywood muscleman, a moderate Republican married to a media celebrity woman from the Kennedy family, the actor known for *The Terminator* and *Kindergarten Cop* movies and for womanizing, had unseated Governor Gray Davis in a 2003 gubernatorial recall election. He promised to rescue the state from the financial mess created by bloated bureaucracy, energy companies misusing energy deregulations, and the revenue downslide that occurred due to the dot com bust in 2000. A fiscal conservative, Schwarzenegger found the state could not sustain its swollen social programs. In his first State of the State address in January 2004, Schwarzenegger harangued the lawmakers, "Every governor proposes moving boxes around to reorganize government. I don't want to move the boxes around; I want to blow them up. We have multiple departments with overlapping responsibilities. I say consolidate them. We have boards and commissions that serve no pressing public need. I say abolish them."[23]

Californians had been very deferential to Schwarzenegger's political and economic agenda, until when he hit the budget impasse and said that the legislators were part of a political system that is "out of shape, that is out of date, that is out of touch and that is out of control." Publically demonstrating his disgust and impatience, Schwarzenegger told an admiring audience in a shopping mall plaza, "If they don't have the guts to come up here in front of you and say, 'I don't want to represent you, I want to represent those special interests, the unions, the trial lawyers . . . ' If they don't have the guts, I call them girlie men."[24] The "girlie men" wisecrack reminisced a satirical skit from an old *Saturday Night Live*, in which two stand-up comedians stuffed themselves as bodybuilding musclemen and mocked Schwarzenegger's Austrian guttural accent and called everybody "girlie men."

That was long before Schwarzenegger ever thought of entering the California political field, but the he-man who was mocked and parodied turned the satire upside down and used the comic-derogatory remark against his opponents with devastating politic effect. His stubborn refusal to apologize for his egregious remarks, referring to his opponents as sissies, raised a scream of protests, including from gay and lesbian groups who had been struggling for equal protection and equal marital rights. Schwarzenegger had never been known for his gender sensitivity. During his election campaign, he was accused of having groped women in the past, for which he apologized for political reasons.

But the episode reflected something deeper about American society. Despite the remarkable progress women have made in many fields and the increasing tolerance being shown toward gays, American society is still a male-dominated, chauvinist society where the greatest insult to a man is to be called a wimp, a sissy, a girlie boy. Call an American a "he-man" or "son of a gun," he feels complimented. Call him a "girlie man" and that could be a cause for a gunfight.

But what could be more outrageous than comparing a politician to a woman's anatomical part, as Hollywood's comedy queen Whoopi Goldberg did about George W. Bush with a bottle of wine and double entendres at a multi-million-dollar gala fund-raising dinner for the John Kerry presidential campaign at Radio City Music Hall in New York in 2004! Obscene sexual jokes have been part of the American public discourse, and the U.S. Supreme Court affirmed their value as free speech in the *Hustler* magazine case, in which, as mentioned earlier, the pornographer had parodied the morally pompous Baptist preacher Falwell as having had his first-time sexual encounter with his own mother in an outhouse. The Court said, though outrageous, let speech be free. But as an example of how free speech and free marketplace try to correct each others' excesses, Goldberg's "bushwhooping" sexual puns did not amuse Slim-Fast, the maker of diet aids. Slim-Fast cancelled its campaign with Goldberg as its spokesperson for the commercial campaign with a catchy slogan, "I'm a big loser . . . if I can be a big loser, there's nothing stopping you."[25]

Goldberg nevertheless was unrepentant and tried to put up a brave face, saying, "Just because I'm no longer in those spots, it doesn't mean I will stop talking. While I can appreciate what the Slim-Fast people need to do in order to protect their business, I must also do what I need to do as an artist, as a writer and as an American—not to mention as a comic."[26] As is the case with most of corporate America, Slim-Fast supports both Democratic and Republican presidential candidates and would rather stay out of trouble, as Disney did by refusing to distribute Michael Moore's anti-Bush documentary *Fah-*

renheit 9/11, which a few years ago turned out to be a much-talked-about box-office hit. Ironically, Goldberg lost Slim-Fast, John Kerry the White House! Whoopi Goldberg was not the only one with her comedic tongue flapping out of control at the Kerry Democratic fund-raising dinner. John Mellencamp, an American rock singer-songwriter known for heartland rock, sang "Texas Bandido," referring to George W. Bush, as "he's another cheap thug that sacrifices our young . . . You're going to get us killed with your lies."[27] Jeering at Bush's deep Christian values, actress Meryl Streep (*The Iron Lady*, *The Manchurian Candidate*) asked the audience if Jesus would have approved "megaton bombs dropped on the sleeping families of Baghdad."[28] Indeed, the Kerry campaign dutifully disapproved the remarks as inapposite and having "crossed the line," but they also asserted that artists and comedians have their freedom. Of course, when they raise millions of dollars, pop culture icons may say anything though they may not help the campaign or advance a candidate's political agenda. Free speech does not guarantee sound political judgment. Raising and spending money during elections is a form of First Amendment protected freedom, as the U.S. Supreme Court ruled in *Citizens United*, but can First Amendment freedoms protect the United States from the patterns of history and the oft-predicted decline and fall?[29] Yes, indeed, it can, as we shall see.

THE FIRST AMENDMENT IS MORE THAN FREE SPEECH

Paul M. Kennedy argued in *The Rise and Fall of the Great Powers* that the global economic balance of power has already begun to move away from the Soviet Union (Russia) and the United States as well as away from the European Economic Community toward Japan and China. He predicted the inevitable rise of the Pacific region, though he did not foresee the economic stagnation of Japan.[30] The United States and Russia, on the other hand, would decline and fall due to military overstretch, the escalating cost of the arms race, and the consequent fall in investment in productive economic activities. The military overreach unsupported by economic growth was the cause of the decline of great European power from 1500 through the twentieth century, including the Hapsburgs, France, and Great Britain. Looking at the cruel patterns of historical currents and the curse of history repeating itself, Kennedy continued his pessimistic predictions about the United States in his subsequent book, *Preparing for the Twenty-First Century*.[31]

The Soviet Union dissipated into the semidemocratic Russia of today, but the United States did not decline. Nor did the United States stagnate. On the contrary, the United States, despite sometimes with gut-wrenching ups and downs, has been growing. For example, in 1987, the year of Kennedy's prediction, the U.S. Gross Domestic Product (GDP) was $4.8 trillion and today it is $15.454 trillion (2012 estimate), according to the U.S. Department of Commerce Bureau of Economic Analysis.[32] GDP is only an economic measure, but let's look at something else. On June 12, 1987, President Ronald Reagan challenged the Soviet leader, "Mr. Gorbachev, open this gate. Mr. Gorbachev, tear down this wall."[33] No one would have imagined at that time that in a little after two decades, in 2009, Americans would tear down their own racial and gender walls and elect an African American, Barack Obama, as their president and Nancy Pelosi as the first woman Speaker of the House, and let a Mormon businessman become a serious contender for the White House in 2012. Instead of declining, America has been evolving and renewing itself, economically and culturally. And if economic and cultural indicators mean anything, the United States, instead of declining, has in fact been growing.

What is the source of America's self-renewal? The system of First Amendment freedoms has created the dynamics of challenge and response that leads to innovative behavior. If history has an irreversible decline-and-fall pattern built into it, the First Amendment has the potential to break the pattern. Consider this, for example: Since the 1980s, America has seen the rise of some of the most innovative corporations that have generated wealth and opportunities not only in the United States but also they have triggered social and cultural innovations in many part of the world. Think about MTV that sprung up in 1981 and whose soft power and cultural impact on the global youth culture, including the Arab-Muslim world, cannot be underestimated. But MTV itself was a technological and cultural innovation, a convergence of two musical streams, one brought by African slaves and the other by Europeans settlers that developed into jazz, country, rock, electronic, hip-hop, rap, and other varieties, the musical forms that had been blending and seeping into each for about two hundred years, creating new, open-ended musical genres. Again, Ted Turner's entrepreneurial vision and successful business venture, the twenty-four-hour news paradigm, TBS-CNN, has been one of the greatest disruptive journalistic innovations of the twentieth century. The new cultural spaces—pop music, MTV, and CNN—were generated by the First Amendment and the aspirational society. Freedom breaks the patterns of history. Free speech creates spaces that never existed before.

Historians Need Lessons

Most amazing has been the rise of Silicon Valley, the cradle of digital revolution—innovators and venture capitalists who built thousands of high-technology companies such as Microsoft, Apple, Intel, Google, Yahoo!, Facebook, e-Bay, and many more; companies that ushered in the information age and transformed the world's sovereign nations into a one-world digital beehive from which no country can escape. And let's not forget another most innovative company, Walmart, whose global supply-chain system, a product of digital revolution and globalization, has impacted millions of people in China, Southeast Asia, and India. Walmart: Always low prices, of course, but in many countries—for example, India—it is also a most innovative farmer, except that it does not do the farming itself but helps the farmer to do better by spreading low-tech innovations. Apple and Walmart, metaphorically speaking, have bound China into a global supply-chain system from which it cannot escape and become a threat to global peace without self-destruction. Historians should learn lessons not only from history and rush to predict the future as a linear regression. They should do "superspeculation"—imagine improbable futures, taking into account the increasing role of science and technology and finance in the age of digital freedom when the whole world is becoming a densely packed matrix of thickly networked nations, as the crisis in the European Union shows. In the networked world no nation can decline, fall, or rise alone today, a phenomenon that poses the greatest challenge to the United States.

In the March/April 2010 issue of *Foreign Affairs*, and recently in his book *Civilization*, Niall Ferguson, a British historian now at Harvard, wrote an article, "Complexity and Collapse," in which he said, "The life cycles of great powers have long been thought to follow a pattern of gradual rise and fall. But what if imperial collapse comes suddenly? A closer reading of history, from ancient Rome to the Soviet Union, suggests that empires fall quickly and without warning. Imperial collapse may come much more suddenly than many historians imagine. A combination of fiscal deficits and military overstretch suggests that the United States may be the next empire on the precipice."[34]

Ferguson uses the framework of complex systems that exist in equilibrium for some time, and then comes the phase change, a sudden transition they cannot handle, and they collapse. He considers the United States an overextended empire, a complex system in that kind of phase change: massive deficit and overstretched military commitment that would lead to sudden collapse. He concludes his essay with a captivating but flawed analogy, "of a runaway steam train that has crashed through the wall of a Victorian railway terminus and hit the street below nose first. A defective brake or a sleeping driver can be all it takes to go over the edge of chaos."[35] Like Paul M. Kennedy and Ian Morris, Ferguson, too, has been caught up in the tyranny of historical patterns and the runaway-train-system imagery of his native Britain, without fully appreciating the source of the strength of American society—the challenge-and-response system of the First Freedoms that has generated the culture of innovation and the aspirational society.

Doubtfully Trusting

Ferguson should have known that Americans have never trusted a single driver. The tripart government of coequals in power subject to the constraints of the Bill of Rights, especially the First Amendment freedoms, discourages the cult of hero-worship. Not only do Americans put the political power seekers through a long, harrowing process of primaries, as we saw during the 2012 Republican primaries; even when they finally elect them and put them into high offices, they keep questioning their ability to govern. They pull them down from their high pedestals, as has been done to John F. Kennedy (Mimi Alford affair), Richard Nixon (Watergate affair), Ronald Reagan (Iran-Contra affair), and Bill Clinton (Monica Lewinsky affair), for example, in spite of their historic achievements. Americans love celebrities and disdain them at the same time. It's difficult for a public person to remain heroic in public consciousness for a long time.

This kind of constant public scrutiny of the system of distributed powers as well as of the people who are given the authority to exercise those powers, whether political, economic, or religious, makes the American system self-aware and sentient. The eternally vigilant, paranoid, self-aware, checks-and-balances system exists in a state of continuous generative turbulence, which has been the source of America's innovation and self-renewal. It is the First Amendment that is the ultimate fount of the aspirational society and the wellspring of the nation's self-renewal.

Great powers of the past were, of course, complex systems, but unlike the United States, they were blind complex systems. They were not self-aware, self-critical, or paranoid enough, so they declined and perished; some gradually, some suddenly. Americans live in a healthy distrust of all institutions and people in power, and nothing sums up better their attitude to authorities than the dictum "Trust, but verify" (Translation of the Russian proverb). Few American presidents have been ever fully trusted regardless of the great work they might have done, and this attitude has done no harm to the United States of America. For example, decades of superhuman efforts by public relations wizards and admiring journalists to transform the Kennedy family into American royalty has culminated into a greater expose of the family members and their sexual escapades.[36] The predisposition to doubt and to question every claim is one of the greatest traits of the American character. Americans are natural-born Doubting Thomases, always unsure, constantly searching and tinkering, the traits that engender innovative behavior.

HOW UNSCRIPTED LIFE ENABLES INNOVATIVE BEHAVIOR

Instead of visualizing the United States as a runaway train, another RMS *Titanic* fatally cruising toward an iceberg, or a latter-day Roman Empire, let's think again. Let's visualize the United States as an incomplete structure, a kind of pyramid in the making, where at its wide-open base there is a diverse range of restless people, social critics from extreme left to extreme right, from Noam Chomsky to Rush Limbaugh, who keep telling us not only that there is something rotten in the state of the United States of America but also who is responsible for the rot and what can be done, and must be done. These people with contrarian dreams keep telling us this is how American should have been, could have been, and ought to be.[37]

And then there is a tribe of utterly crazy people, iconoclasts, humorists, cartoonists, parodists, pornographers, lampoonists, and late-night stand-up comedians, people such as Jon Stewart, Jay Leno, Bill Maher, Steve Colbert, David Letterman, Larry Flynt, Michael Moore, Chris Rock, and let's not forget Lady Gaga—all of these for whom there are no taboos, nothing is sacred, there are no heroes. They are the loud-mouthed idol breakers who puncture the hot air balloons of state- and corporate-created fantasies and propagandas. The continuous discourse of critics, serious, funny, and outrageous, which is born of the First Amendment freedoms, keeps the United States in a state of creative turbulence, at the edge of chaos. They free us from irrational fears and allow the mind to think afresh, roam freely, and be innovative.

Great powers of the past took themselves too seriously; they never laughed at themselves. Americans are a self-derisive people. Americans are always laughing at themselves and their leaders. They know there will be another General John Doe with his pants down, but the U.S. Armed Forces must remain the best ever. There is no hero-worshiping in America. A free society needs no heroes.

It will be naïve to think, however, that there is a direct causal relation between the First Amendment freedoms and innovation.[38] More freedom would not cause a sudden spurt in innovation. Freedom is not an app that any nation can download and use, as Ferguson mistakenly argued in *Civilization.* In fact, freedom is extremely difficult to grasp and handle. It has taken the West more than three hundred years to create institutional structures for freedom to exist meaningfully. India, for example, incorporated a democratic parliamentary-federal system of government, but the country is still struggling how to manage various manifestations of freedom, from free speech to free marketplace.

But let's also keep in mind that innovative behavior is not limited to science and technology; it includes social, political, cultural, and commercial innovations as well.[39] Although there is no causal relation, there is enough empirical evidence available that supports the view that the First Amendment and innovative behavior go together, that there is a strong correlation—especially when you triangulate First Amendment freedoms with two other factors: the free marketplace and the free flow of brain power. Perhaps few other civilizations in the past had this unique, dynamic, triangulated, complex system as in the United States. Theodor Seuss Geisel, for example, would not have become Dr. Seuss but for the system, the triangulated challenge-response system, of the United States of America.[40]

State of Paranoia

At the heart of American society is a set of values, conflicting values, embodied in the Bill of Rights, especially the First Amendment, which by creating a state of controlled turbulence, or ordered liberty, also creates a state of vigilance—in fact, a state of paranoia. Andrew Grove, former CEO of Intel, said a few years ago that paranoia is not unhealthy. In his book *Only the Paranoid Survive*, Grove does not talk about mere survival but turning challenges, "strategic inflection points"—inevitable life-and-death challenges that every company has to overcome—into opportunities for innovation and growth. In the preface, he says how, in the 1980s, Intel was forced out of the memory-chip business by tough Japanese competition.[41] (Japan was on the rise in those days and was expected to dethrone the United States as the world's greatest economic power. China is expected to do the same, very soon. Watch the pundits!) But instead of giving up, Intel plunged into a new field of technology, microprocessors (the building blocks of a computer's intelligence), which in turn revolutionized the whole field of computing. Bill Gates and Paul Allen, the cofounders of Microsoft, built their software empire on the microprocessor, the CPU, the brain of the computer.

Corporations need not only an environment of competition to thrive but also constant public scrutiny, the kind of scrutiny that can come only from free and fearless speech.[42] But when an innovative product or policy is not subject to strict media scrutiny, it can cause havoc, as it happened about Iraq's weapons of mass destruction or Mattel's China-made toys. Or consider Wall Street, where a few years ago several innovative financial products were introduced without intense media scrutiny. Warren Buffett called these complex financial instruments *financial weapons of mass destruction*.[43] The news media could have done a better job, which is a dire warning to us that in even a most open society with a robust free press, media can be distracted and blindsided through ignorance of the complex issues or by spending its limited resources on trivial pursuits; for example, celebrity gossip and the sex life of the high and mighty. The need for more of WikiLeaks-like global antiestablishment social media organizations, dedicated to muckraking, investigative reporting, and whistleblowing, couldn't be greater than it is today, a topic of great importance that we will explore later.

HOW THE FIRST AMENDMENT CREATES MONGRELIZED OFFSPRING

Public scrutiny through the news media is only one aspect of the First Amendment. Equally important is the fact that the First Amendment creates an environment of promiscuity for unrelated ideas to cohabitate and breed new mongrelized offspring. Call it hybridization of ideas. That's what innovation is all about.[44] Consider, for example, putting sensors in Nike running shoes that tell an iPod about your performance and what music to play to motivate you to finish your last stretch.[45] Imagine the future when everyone wears sensor-embedded shoes that are networked with a health care facility for sentient robots to keep track of our health, which, of course, might raise questions about privacy and pose challenges to First Amendment experts to come up with a corrective.

Imagine how the First Amendment created space for Shawn Fanning (and Sean Parker) of Northeastern University to innovate and develop Napster in 1999 to exchange peer-to-peer music files through a digital exchange system.[46] All the building blocks, including the MP3 compression technology, file sharing, and exchange technology, already existed. Fanning and Parker dreamed up something new, but their right of free speech, listening and sharing, as innovative behavior came into conflict with the intellectual property rights of a most powerful multi-billion-dollar music industry, including publishers, music bands, and individual artists. After a long series of not only lawsuits against Napster by Metallica and others but also thousands of cease-and-desist orders issued to universities and individuals, Napster closed the service and eventually went into liquidation in 2001.

But peer-to-peer technology was out in cyberspace, and the recording industry knew that the business model of packaging songs according to their own marketing calculus, in CDs or DVDs, would not survive, especially when consumers, young consumers in particular, had tasted the freedom of choice and how they could cherry-pick and make their song collections on the Internet and share it with others. Technology-enabled expressive behavior is the hallmark of the digital age.

Steve Jobs of Apple understood the threat to the music industry and saw a great opportunity for making music sharing legitimate and affordable.[47] He persuaded major labels and publishers to license music to Apple's iTunes Store to enable people to download a song for ninety-nine cents to their computers or to the iPod, the mobile device that Apple had developed earlier. And we know the rest of the story of recombinant innovation and how this little device has changed the world of popular culture as well as spawned newer smart mobile devices. Napster may be credited with having instigated it. Joseph Schumpeter popularized the concept *creative destruction*, which according to him is a force unleashed by innovative entrepreneurs even though in the process it causes destruction; for example, the turbulence created by Napster that eventually led to its own destruction and the subsequent creation of iTunes and the iPod.[48]

But entrepreneurial creative destruction needs "a zone of unregulated behavior."[49] The source of the unregulated behavior is the First Amendment, which works at several levels. It questions state-created fantasies, myths, orthodoxies, and the cult of hero-worship. The shelf life of a hero in America is very short. Like Tina Turner in *Mad Max Beyond Thunderdome*, Americans seem to be saying, "We don't need another hero."[50] There is a diffused suspicion in American society about the legitimacy of power, the people in power. Power and freedom constitute a zero-sum game; you can increase one at the cost of the other.

Creative Disequilibrium

The First Amendment is part of the Bill of Rights, and all the rights exist in dynamic equilibrium. Although the language of each amendment gives us a feeling of absoluteness, in actual practice the rights limit each other, especially when they come into conflict with each other, as happened in the case of Napster, or when free speech touches privacy, reputation, the criminal justice system, domestic peace, or national security. The fundamental rights are variable values that constantly assert their supremacy, constitutive of and constituting the dynamic equilibrium system.

Regarding the First Amendment, the framers used the language of denial, prohibition, and negativity. Normally we ask the government to do this or that for us, but here were a people, perhaps the first people in the history of mankind, who commanded their government not to do something. That's why the First Amendment language sounds so absolute and so powerful, reminiscent of the Ten Commandments: Thou shall not covet. "Congress shall make no law respecting an establishment of religion, or prohibiting the free exercise thereof; or abridging the freedom of speech, or of the press; or the right of the people peaceably to assemble, and to petition the Government for a redress of grievances."

I came to America with a sense of wonderment, and I can't stop wondering: Who were these founders of the Republic, the people who had just finished fighting and defeating one of the most powerful European powers and then sat down amid raging uncertainties and wrote a power-limiting constitution, including a set of fundamental rights that are not merely directives but are justiciable? It's amazing that the American Revolution did not devour its own children, unlike what happened during the French Revolution.

The interpretation of the First Amendment has been changing. Fortunately, the framers of the Constitution did not leave any explanatory notes about any of the amendments, a fact that has given each generation via the Supreme

Court, the guardian and the final authority on the meaning of the Constitution, to interpret the rights according to the prevailing national mode of consciousness.

Broadly speaking, the founders, by enacting the Bill of Rights, especially incorporating the First Amendment freedoms, were shuttering the door on Europeans tyrannies including the Spanish Inquisition, the English Star Chamber, universal censorship including punishment for publishing without a license, restraint on religious practices, and other tribulations. Each amendment, while affirming a right as a natural right, was also the condemnation of a corresponding European tyranny. The Sixth Amendment, for example, about the openness and transparency of criminal trials by a jury of one's peers, was an indirect condemnation of the special court, the Star Chamber, the English court of law whose sessions were held in secret without any indictment, right of appeal, juries, and witnesses.

Of course, the founding fathers were heirs to the best in European tradition; they were the children of Enlightenment, inspired people drunk deep on the idea of liberty as advocated by Milton's *Areopagitica*, Locke's *Treatises of Government*, and Jean-Jacques Rousseau's *Social Contract*, among others. They shut the door on European tyrannies, but they kept open the doors of Enlightenment. The Enlightenment in a nutshell was nothing but going back to and recovering the rational spirit of the ancient civilizations of Rome and Greece, where reason was the primary source for discovery of knowledge and the legitimacy of power. The Greeks, for example, having consigned their gods to Mount Olympus, were free to think rationally and argue fiercely about all human affairs. So has been the case with the Americans, who, having separated the church and the state, relegated the church to the status of any other corporation subject to public scrutiny. No wonder it is only in America that people have been asking the Catholic Church regarding sexual abuses: What did the pope know and when did he know? The same question that they asked Richard Nixon and Enron's CEO, for example.

Warning

One is likely to be carried away, but here is the irony. Having created an immortal document in 1791, the Bill of Rights, and having assured themselves of the freedom of speech and the press, the first few generations of free Americans totally ignored it, more or less. In 1798 the Federalists in the Fifth U.S. Congress enacted a most horrendous anti–free speech and press law, the Alien and Sedition Acts, that made it a crime to publish "false, scandalous, and malicious writing" against the federal government or its officials or the official acts.[51] The law was enforced to punish the opponents of the Federalist-controlled government of John Adams, including some private citizens and about two-dozen editors. In 1801 when Thomas Jefferson, a

Democratic-Republican, became the president he pardoned everyone convicted under the Act, which in any case expired on its own due to the sunset clause. Many political leaders, including James Madison, who is credited with crafting the First Amendment, and Thomas Jefferson protested that the law violated the First Amendment freedom because it punished the discussion of the government policy.

By today's standards it was a gross violation of the law, but its constitutionality was not tested in the U.S. Supreme Court because due process was not well established by that time. Under due process, the courts rather than government or legislators describe, delineate, and determine as to what constitutes justice, liberty, and fundamental fairness in society. What was John Adams afraid of? The Federalists did not want the United States, the newly founded but yet-to-be-stabilized nation, to get unnecessarily involved in Napoleon's European Wars (1803–1815); nonetheless, they favored England because of the foreign policy dispute with France, who had demanded an American war loan and other concessions, which incensed Americans when they learned the details of the diplomatic negotiations. Jeffersonian Republicans, on the other hand, were somewhat sympathetic with France, keeping in mind how the French had helped the Americans during the War of Independence or the American Revolution. To silence the critics of John Adams's foreign policy and war preparation efforts against France, who had been attacking American shipping, the Federalist Congress passed the Alien and Sedition Acts, which the public supported due to the dread and revulsion of the French revolutionary doctrine as well as the fear of French immigrants.[52] Americans have been always cautious about immigrants in spite of the fact that newcomers have been constantly replenishing America's muscle and brainpower.

The Alien and Sedition Acts violated the First Amendment. But laws are worthless unless both the government and the people obey them. Rights have no meaning unless we value them with our hearts and souls and are ready to fight for them, and there is an independent institution that supports and enforces those rights. In 1776 the American people proclaimed a charter, the Declaration of Independence, and subsequently by attaching the Bill of Rights as a rider to the Constitution in 1791, they gave it a concrete, legally enforceable form. But the American people forgot their rights. Or they did not know where to go for redress, especially when the government itself was the violator of their rights.

The nineteenth century was a century of nation building for Americans: the Louisiana Purchase from France in 1803 and the westward expansion; the long preparation for the Civil War in the theater of the public mind until it actually began 1861; and the Reconstruction. Equally important was the development of new communications technologies—the telegraph, telephone, sound recording, radio, photography, and the motion picture—and the offset printing press that gave rise to newer forms of journalism such as investigative, sensational, and tabloid journalism, as a successor to the Penny Press.[53] The Penny Press, easy-to-read newspapers with human-interest stories, itself was a great cultural innovation of the early nineteenth century, as mentioned earlier.

But the most important change that occurred after the Civil War was the adoption of the Fourteenth Amendment to the Constitution, which extended the Bill of Rights to all states. The Fourteenth Amendment not only made the states' sovereignty incidental but also the due process clause required states to follow certain procedures before a citizen was deprived of his life, liberty, or property. The equal protection clause required states to provide equal protection under the law to all citizens under their jurisdiction. A Yale historian, David Blight, has called the Fourteenth Amendment "a revolution that revolutionized the Great American Revolution."[54] It was a tipping point, a phase change. America could never go back to its past. But here is the irony. For more than a hundred years, in spite of the First Amendment and the Fourteenth Amendment, the federal government and state governments continued messing with people's free-speech rights.

WHEN THE U.S. SUPREME COURT WOKE UP

The U.S. Supreme Court had not begun to assert its role as the guardian of the people's civil liberties until the time of World War I. In order to make the world "safe for democracy," on April 6, 1917, Congress approved President Woodrow Wilson's request to declare war against Germany, and the same year he enacted the Espionage Act of 1917. The Act criminalized any attempt to obstruct military operations, assist America's enemies during wartime, encourage insubordination in the military, or impede military recruitment and war efforts. The Act was amended and extended a year after (Sedition Act, 1918) to include "any disloyal, profane, scurrilous, or abusive language about the form of government of the United States . . . or the flag of the United States, or the uniform of the Army or Navy."[55] Just as the Alien and Sedition Acts of 1798 were passed to save the United States from the dangers of the French Revolution, the Espionage Act of 1917 was passed to empower the government to curtail free speech in order to make the world "safe for democracy."

Apart from danger from the Kaiser's Germany and the fear of the so-called enemy aliens of German origin (of which the Seuss-Geisel family of Springfield, Massachusetts, for example, was one), there was also danger from Communist Russia under Lenin and Stalin. And Americans felt threatened not only by the Germans and the Russians amid them but also by the Soviet Union and the revolutionary motto: "Workers of the world, unite."

Communism challenged the capitalist system and the ruling classes everywhere. Whenever Americans perceive an existential danger, their hearts and minds begin to close and freedom begins to shrink, as it happened during World War I, World War II, the McCarthy era, and after 9/11, for example. Perhaps that's true of every society: survival before justice and freedom.

Between the Espionage Act of 1917 and the PATRIOT Act of 2001, between Thomas Woodrow Wilson and George Walker Bush, comes the U.S. Supreme Court, not necessarily the last best hope for the American people; nevertheless, it is an institution that no one can ignore, as Al Gore discovered in December 2000 and as we are discovering now regarding its decisions about campaign finance, corporate speech, and the right to marry. Landmark decisions such as *Brown v. Board of Education* (1954) and *Roe v. Wade* (1973) have shaped American society as no president or Congress has ever done, except Abraham Lincoln.

The case that aroused the U.S. Supreme Court from its apathy and long slumber was *Schenck v. United States* (1919). Charles Schenck, secretary of the Socialist Party, distributed leaflets to eligible draftees in violation of the Espionage Act, urging them not to submit to intimidation and to assert their rights and arguing that military conscription amounted to involuntary servitude and that is slavery banned under the Thirteenth Amendment. But the U.S. Supreme Court did not buy the argument and, considering the fearful international situation, the Court unanimously upheld the conviction, establishing not only a category of exclusion for free speech but also, more importantly, establishing a standard called the clear and present danger test for determining socially harmful speech. Justice Oliver Wendell Holmes wrote the unanimous opinion, explaining the constitutional standard and the reasoning behind it. He wrote: "The question in every case is whether the words used are used in such circumstances and are of such a nature as to create a clear and present danger that they will bring about the substantive evils that Congress has a right to prevent."[56]

The First World War was over, but the threat of an international communist movement persisted. Under the Espionage Acts hundreds of people were convicted. There were several appellate challenges, and some of them reached the Supreme Court. Charlotte Anita Whitney, who came from a distinguished California family, was a suffragist and a founding member of the Communist Labor Party. She was charged with advocacy and violent overthrow of the government. In a 1927 case, *Whitney v. California*, the U.S. Supreme Court unanimously upheld her conviction based on a dubious doctrine called "a bad tendency test," or one might say, a probable danger test.

FREEDOM FROM IRRATIONAL FEAR

Justice Louis Brandeis, although he agreed with the Court's unanimous decision in the *Whitney* case, argued in his separate concurring opinion, in which he was joined by Justice Oliver Wendell Holmes, that the speech to be excluded from the First Amendment protection must be such that it not only creates a clear and present danger—it raises an imminent threat of violence. Mere advocacy of violence is not enough for speech to be punished. The court majority did not agree with their dissent; nonetheless, in the course of time their views finally prevailed.[57] As Justice Brandeis wrote in his dissent: "Fear of serious injury cannot alone justify suppression of free speech and assembly. Men feared witches and burned women. *It is the function of speech to free men from the bondage of irrational fears.*"[58] (Italics added)

Since the time of ancient Athens, speech has been deemed as an instrument for discovering and confronting truth, creating social good through advocacy and persuasion, civic participation, submission to the gods, and also one might add as an instrument of fear and torture. But this was the first time anyone said so explicitly that "*the function of speech is to free men from the bondage of irrational fears.*" Free speech as an antidote to irrational fear was a revolutionary idea and, as we shall explore in this book, it is indispensable to the formation of the culture of innovation and the aspirational society.

"To justify suppression of free speech," Justice Brandeis continued, "there must be reasonable ground to fear that serious evil will result if free speech is practiced. There must be reasonable ground to believe that the danger apprehended is imminent . . . Only an emergency can justify repression."[59] But nothing should be done except through due process, guaranteed by the Fifth and Fourteenth amendments. Due process makes the government subservient to the rule of law. The Supreme Court is the ultimate guardian of due process. Due process is a most sacred, secular doctrine in the United States.

"The incitement of violence standard" for limiting free speech, as it has evolved in the course of time, has been applied to myriad cases, including newspapers publishing stolen documents from the CIA, also known as the Pentagon Papers; publishing information about how to make a hydrogen bomb; access to information during war time; personal injury cases against the media; for example, a teenager listening to the Ozzy Osbourne lyric "Suicide Solution" and then committing suicide; harm through imitation; for example, a guest at late-night comedian Johnny Carson's show performing a stunt that was copycatted by a teenager who hurt himself; and not least, the rehabilitation of commercial speech as a fundamental right as well as a necessity for economic growth and the functioning of the free market.[60] In each instance the Supreme Court considered whether the danger present in the speech-act situation caused harm or rose to the level of the incitement of violence, and the Court ruled, no, it did not. In free speech cases, the burden of proof of harm lies on the plaintiff whether it is the government, a corporation, or an individual. The courts, not the legislature or the government, will decide after due process how serious the danger is that free speech must be restrained or punished.

Tyranny of Self-Censorship

Various court decisions, including a 1964 landmark decision, *New York Times v. Sullivan*, finally established the primacy of free speech in the American public life. By focusing on freedom of speech and of the press, the U.S. Supreme Court freed the American media from the tyranny of self-censorship. The broadened freedom has enabled the media to create a plethora of new entertainment programs that not only enrich our popular culture but also directly feed the global marketplace through the expansion of commercial speech rights, as we shall discuss later.

Free media and free marketplace form a loop where all innovations are tested. Remember the old advertising slogan "Where is the beef?" If you have an innovative product or an idea, one would say "let's test it in the marketplace." But trust in the marketplace depends upon how a product or an idea stands the scrutiny of the media, which, as Justice William J. Brennan said in the *Sullivan* case, must be "robust, uninhibited and wide-open."[61] Even falsehood and outrageous speech might have to be tolerated for truth to emerge, especially in regard to public affairs. That's the only way to get rid of our irrational fears. That's what the U.S. Supreme Court has been trying to do since the early twentieth century: freeing the American people from the bondage of irrational fears by asking them to test their ideas in the marketplace. Consider the marketplace as a platform for exchange, negotiation, bargaining, testing, and trust building. The marketplace, like the Greek Agora or the Indian Bazaar, is also a site for civilization.

The First Amendment's nexus with the marketplace and America's access to global brainpower through immigration and technology, the power of triangulation, has given American society a distinctive, innovative character. The First Amendment has turned America into a permanent beta state—beta as in software—full of bugs and imperfections for the people to improvise and correct as they go along. Failure is no shame in the United States. People don't commit suicide because they fail, as they do in some other countries. America is a country of second chances. Consider Eliot Spitzer, the disgraced ex-governor of New York, who was given a second chance to redeem himself in the public eye as a CNN commentator in 2010. I wonder if any other society would have given him another chance. To paraphrase Oscar Wilde, every sinner has a future in America. The marketplace, however, decided that Spitzer had little worth in him, and his show *In the Arena*, was canceled only after nine months.[62] But Bill Clinton has fully rehabilitated himself and is a force for good on the global stage.

Failure is not only tolerated but also America's liberal bankruptcy laws fuel entrepreneurial creative destruction. The First Amendment has enabled Americans to develop into a nation of tinkerers and garage innovators. Consider *garage* as a metaphor for *autonomy*, one's own space; it could be a shed, an attic, dorm room, or hacker-space. For most teenagers today it's the laptop, tablet, or smart phone that has become the proverbial garage. Everyone is looking for a new game changer, a new killer app.

The First Amendment has created a mechanism for self-correction through a continuous stream of critical social discourse in the media, which by and large have been bold and unconstrained. The bigger a public figure or a corporation, the bigger the public scrutiny: Richard Nixon, Bill Clinton, Newt Gingrich, John Edwards, Arnold Schwarzenegger, Enron or Walmart, the Catholic Church, it does not matter: everyone is accountable in the aspirational society. The First Amendment frees the mind to explore "the adjacent possible," to search and seek solutions wherever it can. It liberates the mind to accommodate the idea of disobedience, rule breaking, and transgression, which leads to innovation.

Last but not least, the First Amendment has given Barack Obama the wings of hope and Tea Partiers' the audacity to pluck growing feathers from Caesar's wings.[63] This kind of dynamics, turbulence, and harmony is the cradle of innovation, creativity, and self-renewal, the making of the aspirational society, as we shall explore further.

Chapter Two

How Free Speech Stimulates Innovative Behavior

THE REMAKING OF MOTHERHOOD AND OTHER WONDERS

Our biotechnology-based future need not necessarily be posthuman in the sense Francis Fukuyama has dreaded: "to lose our humanity—that is, some essential quality that has always underpinned our sense of who we are and where we are going." This is because humanity's collective cultural heritage of compassion, morality, ethics, and humanitarianism, built over thousands of years on the foundation of major religious faiths, will not disappear because of scientific advances in human genomes.[1] On the contrary, globalization has enhanced our sense of humanity, the need to protect the environment, the urgency to fight hunger and disease all over the world. There was never a time in history when human rights had become such a dominant, universal passion as they are today. Governments, corporations, research establishments, and social organizations define themselves by their human rights policies and actions.

A few months into his first presidency, President Barack Obama issued an executive order lifting all restrictions on stem cell research and explained that it was necessary to do so because

> Medical miracles do not happen simply by accident. They result from painstaking and costly research—from years of lonely trial and error, much of which never bears fruit—and from a government willing to support that work. From life-saving vaccines, to pioneering cancer treatments, to the sequencing of the human genome—that is the story of scientific progress in America. When government fails to make these investments, opportunities are missed. Promising avenues go unexplored. Some of our best scientists leave for other

> countries that will sponsor their work. And those countries may surge ahead of ours in the advances that transform our lives.[2]

President Obama's rationale, however, missed a most important point. Although scientific and technological progress in the United States has been immensely benefiting from the investment made by the government in infrastructure as well as the direct funding of research, this is also equally true of other countries. What makes the United States unique is the endless space created by the First Amendment freedoms that not only attract talented scientists from all over the world but also enable ordinary people to take risks and experiment with their lives unconstrained by the orthodoxy of religious beliefs. Consider this, for example: In 2001 the New Jersey Supreme Court decided what at that time seemed to be an unusual case: a divorced couple's dispute regarding the ownership of a jointly created unique asset, embryos.[3]

Thanks to constitutionally protected property rights and venturesome marketing of reproductive technology, a simple tale became so complicated. A woman named JB married a man named MB in February 1992. Unfortunately, she had a miscarriage, and due to some medical condition that she might have developed during her childbearing complications, JB found it difficult to conceive again. Since the husband did not have any procreative problems, the couple decided to try other means and went to the Cooper Center for In Vitro Fertilization in New Jersey and did a successful in-vitro fertilization procedure. Out of eleven embryos developed at the Center, four were implanted in JB, and one of them resulted in the birth of their daughter. The remaining seven unused embryos were preserved at the Center in case the couple decided to have more children in the future. The couple had to sign a consent form with the Center that in the eventuality of the breakup of their marriage, the court would decide about the fate of the remaining embryos. This is not a posthuman condition but an example of people exercising their natural-born right to life, liberty, and the pursuit of happiness within the confines of the law.

In America today marriage is a delicate and fragile bond. Couples grow up, differentiate, and love disappears. Divorces and separations are not uncommon, and it happened in this case, too. Six months after the birth of their in-vitro child, the couple separated and subsequently filed for divorce. The woman asked in her divorce claim that the frozen embryos be destroyed, but the man disputed the claim and asserted that the embryos also belonged to him. He had the right to use them in the future to have children with another woman. Although the custody battle for children makes divorce a nightmarish experience for some people, lawyers somehow do manage to find ways of dividing the material possessions and financial assets of divorcing couples. But in the case of the New Jersey couple, the battle was over the unused frozen embryos in the custody of the Cooper Center. Children of divorced parents grow up and eventually go away regardless of the initial custody battles. Property can be sold and proceeds divided equitably in court or in an out-of-court settlement. But the JB-MB case was unprecedented: what's to be done with the frozen embryos? They could not be disposed of as property. Although many Americans believe that life begins at conception and an embryo is a potential life, IVF clinics usually destroy them, sometimes without the permission of the couple if they don't claim them after a few years. It is out of these unused embryos that embryonic stem cells have been developed. In 2001, President George W. Bush used his executive power to authorize the federal government, with restrictions, to fund sixty stem cell lines for biomedical research purposes to find cures for such diseases as Alzheimer's, Parkinson's, and diabetes, but President Barack Obama, as mentioned earlier, soon after he took office removed all restrictions on stem cell research.[4]

Right of Procreation

The New Jersey Supreme Court, affirming the decision of the two lower courts, ruled that the frozen embryos of the couple could not be used for implantation in another woman for producing a child without the woman's (JB's) consent because that is tantamount to compelling her to become a biological mother against her will. A woman shouldn't be forced to become a mother because motherhood is an act of choice, the Court said, keeping in mind the ramifications of *Roe v. Wade* in which the U.S. Supreme Court ruled that under the Fourteenth Amendment due process clause, a woman's privacy rights included the right to abortion, balancing the viability of prenatal life and the need to protect the woman's health.[5] The originator of the fertilized egg, not the surrogate womb, determines the rights of motherhood. Justice Deborah Poritz, writing for the unanimous Court, said that the woman's "fundamental right not to procreate is irrevocably extinguished if a surrogate mother bears (her) child. We will not force JB to become a biological parent against her will."[6] Besides, she added, "MB is already a father and is able to become a father to additional children, whether through natural procreation or further in-vitro fertilization. In contrast, JB's right not to procreate may be lost through attempted use or through donation of the pre-embryos."[7]

Imagine the future when a fertilized egg or pre-embryo could be implanted in an artificial womb, and that day may not be far off, keeping in mind medical innovations unconstrained by religious orthodoxy. In that case the real mother would unquestionably be the woman who gave the egg. But could a divorced woman ask that the pre-embryos, which she developed with her former husband, be destroyed so that no one could ever use them for not only procreation but also for stem cell research? The man has his rights of paternity, too, so the New Jersey Court allowed MB to have the embryos preserved in a cryogenic state indefinitely, provided he could afford to pay the fees. In that eventuality, if JB, the woman in the dispute died, the right to the use of the embryos would revert to her ex-husband, MB, and he could use the frozen embryos for implantation and have children. Or in the changing mode of consciousness and the broadening of the idea of marriage and family, MB could donate the leftover frozen embryos to a gay couple to enable them to raise a family. The First Amendment creates new cultural spaces and endless possibilities. On the other hand, JB, ironically, in spite of her unwillingness to procreate, would become a mother posthumously. Posthumous motherhood may become another landmark case for the U.S. Supreme Court to decide.

Mitigating Reproductive Constraints

This kind of dialogue about innovation in social and biological relations continues in the United States, thanks to free speech that makes the articulation of human possibilities, including motherhood, any which way possible. The human reproductive technology has now advanced to a stage where a woman's eggs can be cryogenically preserved to enable her to have a child at any age, even after menopause. This does not make menopause an insignificant event, but there is on the horizon a new threshold for women—freedom from the reproductive constraints that menopause imposes on them.

In 1997, a team of researchers at Reproductive Biology Associates in Atlanta, Georgia, first successfully froze and later thawed the eggs of a twenty-nine-year old woman without damaging the gene-carrying chromosomes.[8] Next, the donated eggs were fertilized with the sperm of the man whose wife—the thirty-nine-year old recipient woman who had become prematurely menopausal but had a healthy womb and wanted to have children—had the embryos implanted in her. She gave birth to a healthy twin. If the egg-freezing technology was available earlier, the woman might have cryogenically frozen her own eggs for later use when menopause disabled her from egg production. Elizabeth Cohen, senior medical correspondent for CNN, reported that according to Dr. Daniel Shapiro, medical director of the egg bank at Reproductive Biology Associates in Atlanta, "Freezing eggs for one's own future use has become a growing trend, with an estimated 5,000 women nationwide making use of the technology."[9]

Today, a young woman on a long academic and career ladder might postpone marriage and child bearing by preserving her frozen eggs for the future when her eggs won't be so healthy. She may even donate her young and healthy frozen eggs to infertile women. Since the quality of eggs deteriorates as a woman ages, chances of babies with physical defects and mental retardation, according to medical experts, increase when a woman in her forties or in later years gives birth to a child. But now an older woman with a secure savings account of younger frozen eggs could beat the odds and become the mother of healthy babies. Frozen sperm banks have been in existence for a long time, and freezing embryos has given no problem to doctors, as we saw in the JB-MB case. Freezing and thawing ova could have damaged them, but by injecting the sperm directly into the thawed eggs through a method called *intracytoplasmic sperm injection* (ICSI), the Atlanta researchers succeeded in saving them.[10]

There may be less ethical objection to the freezing of ova rather than embryos, which are fertilized eggs ready to develop into a fetus and blossom into full life in the womb.[11] In the twenty-first century, a single woman in her fifties could go to a reproductive bank in a fertility clinic and choose the cryogenically preserved eggs of a twenty-one-year-old MIT student and have them fertilized with the sperm of a musician, lawyer, or an athlete, and let her womb do the rest, if it can.[12] The woman still has to undergo (and perhaps enjoy) the pain of pregnancy, but her menopause or infertility would be no handicap to having children. Freedom from the chains of menopause and reproductive equality with men, who can have children at any age, are some of the most important gains of the new biotechnological development, which would not have been possible without the freedom from religious constraints guaranteed by the First Amendment and the Fourteenth Amendment. Thus, biotechnology enhances human rights.

Seeking Answers

The First Amendment does not answer all our questions, but it liberates the mind to seek answers wherever it can because it frees us from irrational fears. The question of whether the womb of a fifty-year-old woman can be made as good as that of a twenty-one-year-old has yet to be answered, but the answer does not lie with the state—it lies with the bold tinkerer, the experimenter who imagines all possibilities, including the consequences. If women begin to depend on the promise of the later-day pregnancy offered by the frozen-egg technology, they may not be able to raise their children as well as would be done by a young mother: the mom who goes to every Parent Teacher Association (PTA) meeting and raises hell if the school cafeteria does not serve a healthy, organic lunch or the library keeps objectionable reading material, for example.

Every time a new technology is developed in someone's laptop, garage, or laboratory, it gives us a sense of liberation and power over our destinies. We feel that our free will has been enhanced, and we could make informed decisions and better choices. But along with free will we need wisdom to ponder over the unintended consequences of the new regenerative freedom, which is ultimately a social platform to be tested and contested and can be done so only if the press is free, robust, and strong. Without freedom of the press we will have a posthuman brave new world of genetically engineered, gender-selected baby farms, perhaps a Fukuyama's nightmarish scenario. The First Amendment lets randomness play its role in social evolution, and consequently the patterns of history are broken and the future remains unpredictable and undetermined.

Although the government has a substantial role to play in invigorating scientific research, as President Obama mentioned in promoting his stem cell research policy, it is the First Amendment that stimulates innovation in American society by challenging the innovator to go and test their claim in the competition of the marketplace under the watchful eye of the news media. The marketplace, however, should be considered in its larger and more encompassing metaphorical sense of a dynamic platform for exchange of any kind—of ideas, goods and services. An innovation is meaningful only if it is functional and socially useful—in the sense that it can be tested in the marketplace, can be monetized, increases efficiencies, and creates future possibilities. Consider some examples: the inverted pyramid narrative form of storytelling that developed as a response to the telegraph and consequently turned journalism into a medium for the masses; today's collaborative brain-sharing through outsourcing that has enhanced research, manufacturing, and services; social media whistleblowing enterprises such as WikiLeaks that keep the powerful in check; the U.S. Supreme Court's socially transformative interpretation of the First Amendment free press clause giving it a preferred position in American society that has broadened our freedoms; and Franklin D. Roosevelt's Social Security program that has not only made the lives of the elderly livable but also pumps billions of dollars annually into the economy, stimulating economic growth.

The news media not only stimulate the diffusion process of an innovation but also provide the marketplace an unbridled source of critical feedback from the public, which acts as a further incentive for the innovator to improve the innovation. In a challenge-response environment created by the free press, every innovation, every killer app, becomes a fodder for the next one, disrupting the status quo and pushing the society once again to the edge of chaos and creative destruction. The marketplace for motherhood biotechnology discussed above is at a formative and evolving stage, and the mode of consciousness that would wholeheartedly accept such advances might not be there at present, though it shows us what innovative minds can do. Through their probing and extensive reviews and analyses, both serious and comedic, the media routinize innovations and make them socially normal and profitable for the marketplace, until the next phase change occurs and pushes the society to another cliff, to the edge of chaos where new order begins.

FREEDOM FROM REVEALED TRUTH

The American innovator is the offspring of a cultural environment that has been evolving for more than two centuries, an ecosystem predicated upon the least government restraints, rules, and regulations that are absolutely necessary. The First Amendment commands governmental noninterference, but it does not enjoin the government for creating infrastructure to augment innovation culture except through copyright and intellectual property laws.

Most First Amendment scholarship is limited to the discussion of free expression for political and social reasons; for example, participation in democracy, keeping checks on the government, maintaining social stability, and pursuing personal fulfillment. But overlooked is the fact as to how free speech and freedom of the press have enabled generations of Americans to be technologically and socially innovative and solve societal problems in creative ways, and how this process is continuing in the Internet Age. It is important, therefore, to ascertain how innovation is correlated with freedom of speech; how the American core values embodied in the First Amendment instigate and drive competitive innovativeness in the United States; and whether American society can break the pattern of history and escape the paradigm of decline and fall.[13] Decline and fall is inevitable if a society is closed and its collective mind is shackled by a single dominant ideology or religious orthodoxy, where institutions and platforms for critical self-examination and untrammeled public discourse are not tolerated.

The First Amendment leads to diversity of thought that stimulates innovative and outrageous behavior. Through satire and ridicule as late-night iconoclastic comedians do, for example, it prevents any overarching dogma to hold the American mind for too long. The First Amendment has turned America into a perpetual mode of questioning, observation, experimentation, and pragmatic adaptation. In such a society failure is not stigmatized but instead it is measured and analyzed; it adds fuel to the fire of innovation through trial and experimentation. It keeps society in a beta mode, in a state of perpetual equilibrium-disequilibrium, between harmony and turbulence, always looking for something new, some fresh solutions to vexing old problems. Consider this, for example: A few years ago an American woman gave birth to a girl whose faulty genetic system was corrected at the embryonic stage. Without doing genetic tinkering in the embryo, the girl would have gotten the defective gene of her mother, a thirty-year-old woman with the Alzheimer gene, who was destined to become a victim of the dreaded disease. Genes do shape our destiny; nevertheless, now it is possible to eliminate a bad gene and have a different fate.[14] In this kind of posthuman future, there is hope for the hopeless, and that's why everyone wants to do what the American researcher and innovator is doing: look to science and technology to solve the problem of human suffering and make living better for everyone.[15] How is this kind of innovative behavior correlated with First Amendment freedoms? The First Amendment drives people to "to explore strange new worlds, to seek out new life and new civilizations, to boldly go where no man has gone before," and become a perpetual explorer rather than be a blind follower of some revealed sacred truth.[16]

More than a decade ago something extraordinary happened in outer space, but in this age of rapid innovation, the technological feat looks so ordinary now. On March 1, 2002, American tinkerer-innovators went to space to repair the Hubble Space Telescope, the humankind's eye on the cosmos. American astronauts rode the shuttle *Columbia* to rendezvous with the Hubble Space Telescope to replace its aging solar wings and upgrade its other instruments, and after two days of chase at 17,000 mph, they finally caught up with the space observatory 360 miles above the earth over the Pacific Ocean. Using a robotic arm, the astronauts pulled the Hubble Space Telescope to the shuttle's cargo bay area for repair and renovations, much like marine biologists tug a sick whale to shore to nurse it to health and then release it to the ocean. After hours of working in space, the space-walking handymen installed the new wings that gave the Hubble 20 percent more wattage and enabled its instruments, including a new powerful camera and power system, to work more efficiently in order to enable us to observe the universe better.

In spite of the leak in *Columbia*'s cooling system, such an impossible mission became possible because of a daring and adventurous spirit, though, earlier too repairmen had been into space for a tune-up service on the Hubble. This was the fourth Hubble service call and the most daring, though not the last one, that would not have been possible without the culture of aspiration, openness, and critical self-examination nurtured by the First Amendment.[17] The First Amendment instigates civic participation and critical evaluation into public and private organizations' activities and acts as a corrective mechanism. In the culture of the aspirational society, there's no difference between a genetic innovator repairing an embryo to rid an unborn girl of the Alzheimer-causing gene and a space handyman implanting new eyes, heart, and nervous system on the Hubble so that we could understand the universe a little more. Like the Internet, which was originally developed as a decentralized system for defense research and military applications and has spawned an unprecedented era of innovations and globalization, genetic engineering, embryonic technology, and space explorations are the products of the cultural spaces created by First Amendment freedoms. It is only through the fearless freedom of the news media that it is possible to debate various ethical, functional, and safety issues of new technologies, a process that instigates further challenge and response and civic participation in the aspirational society.

ETERNAL OPTIMISM FOR MAKING THE WORLD A BETTER PLACE

What intrigues an outside observer the most is the eternal optimism in the United States that the American innovator will make the world a better place. As Thomas Paine said in *Common Sense*, "We have it in our power to remake the world."[18] The source of this self-renewing aspirational impulse is the dynamic system of the nation's core values, the free marketplace of ideas nurtured by the First Amendment. The energy released by the checks-and-balances structured system of freedom is so overwhelming that it not only overpowers the forces of negativity, it generates confidence that propels people to tinker and innovate. The dynamic system of First Amendment core values, a central concept in the theory of self-organizing and self-renewing systems, defies determinism and breaks the patterns of history.[19]

The First Amendment, a simple but powerful commandment that the government shall leave people alone so that they can organize their energies to work for goals and objectives that they personally value and consider worthwhile, has become a kind of evergreen field of dreams, a perennial source that releases enormous constructive energy for tinkering, creativity, and innovation to be tested in the open and competitive media and marketplace. In several landmark cases the U.S. Supreme Court has treated the First Amendment as an indispensable tool for the advancement of marketing, entertainment, and technological innovations, as we shall discuss in the following pages.[20]

What is an innovative mind-set? When encountering a problem, an innovative mind summons up its energies to come up with new perspectives, designs new devices, develops new programs based on unconventional approaches, and not only solves the problem at hand but also creates associative possibilities.[21] But this can be accomplished only if the society creates large, unregulated spaces by leaving people alone, not as a matter of state policy, but as something essential to being an American. Consequently, the First Amendment's guarantees of governmental noninterference has encouraged fearless speech, investigative reporting, unrestrained thought, and unceasing innovation through endless experimentation leading to newer developments in science, technology, the arts, and social relations.

Most importantly, freedom of the press creates an environment of divergent thinking, nonconformity, and resistance to peer pressures. It allows Americans to be creatively crazy, to be impure, and to mix and remix ideas to create new cultural forms; for example, as mentioned earlier, the evolution of pop music due to the convergence of African slave music and European ballad tradition that gave rise to uniquely American musical forms such as jazz, rock, and hip-hop.[22] The evolution of popular music, the opening up of sports to African Americans, and the Martin Luther King Jr.–led civil rights movement gradually led to transformation in the mode of consciousness and the network of people's assumptions about race, ultimately culminating in the flight of the black swan, Barack Obama, to the White House. Without the First Amendment freedoms, this kind of continuous social and cultural innovation in American society would not have been possible.

Sustainable Innovation

In the past, several societies showed technological and social innovativeness, but they were unable to sustain themselves because they had no self-correcting and self-renewing institutional system based on free and untrammeled thought supported by freedom of the press. After the Bolshevik Revolution of 1917, the establishment of the Soviet Union itself was an innovative system, but without critical self-examination and the corrective mechanism enabled by free speech, the system could not be sustained, though in 1957 the Soviet Union was the first to launch a satellite into orbit that ushered in the space era, and its space station was in full operation at the time of its collapse in 1991.[23] Hitler's Germany had space rocketry, modern electronics, and nuclear energy and atomic bomb development programs in fairly advanced stages, and even if the Allies had not defeated Germany, the system was too closed to have sustained itself for long. Without a robust exchange with the outside world through trade, people, and ideas, Germany would have atrophied or imploded. Many mathematicians, engineers, and physicists had left Germany because of the politicization of German universities, apart from the fact that Jewish scientists and academicians were purged from academia.[24] Japan under Emperor Hirohito, although it had developed a war machine based on a highly efficient industrial economy, possessed one of the largest merchant fleets in the world, and had developed engineering methods that enabled it to manufacture low-cost, high-quality products for export, it was a closed, isolated country surrounded by poor, helpless neighboring countries such as China and Korea, which might have given it an illusion of a great power and invincibility but in fact it was no different from Nazi Germany.[25]

These seemingly indestructible societies lacked the essential institutional system of critical self-examination provided by a free press that enables a living social system to do environmental surveillance and adapt and renew itself through the dynamics of challenge and response. Although communist China today has given up its Leninist-Maoist economic development model and has adopted state capitalism to advance economically, no doubt a bold innovation in itself that has enabled it to make unprecedented economic growth, one wonders if this kind of growth can be sustained for long without the battle of ideas and the internal checks and balances facilitated by free and fearless speech that facilitates a society to make a smooth political transition from one generation to another.

Writing in *Foreign Affairs*, Professor Robert O. Keohane of Princeton University said, "We know that autocracies are fundamentally less stable than democracies. Lacking the rule of law and accepted procedures for leadership transitions, the former are subject to repeated internal political crisis, even though these might play out beneath a unified and stable façade. China's leadership crisis during the spring of 2012, marked by the detention of the politician Bo Xilai and his wife, illustrates this point."[26] How much free speech contributes to making a society continuously and sustainably innovative and renewable is an important question and should be also of great interest to global corporations and nonprofit enterprises.

During the past few decades several scholars and media experts have attempted to explore the sources of creativity and innovation in American society and have made an objective study of the United States as a dominant hegemonic force in international affairs. And depending on the traditional wisdom that what goes up must come down, especially when a society overextends itself, they have come to similar conclusions that the increasing complexity of the networked globalized system would make it difficult for the United States to continue its growth, that its systemic or inner sources of self-renewing energy would dry up. In spite of the fact that America is "the greatest engine of innovation that has ever existed," the country is losing its edge because the resources are being frittered away in unproductive adventures, and investment is not being made in education and infrastructure.[27]

Decline-and-fall historians and public intellectuals, who see deterministic patterns in the march of history, attribute the U.S. global preeminence to historical advantages created by geography-and-resource-based engines of economic growth rather than something more fundamental, which is the free marketplace of ideas, a core value embedded in the U.S. Constitution that has led to the rise of phenomenal corporations such as Walmart, Apple, Microsoft, Facebook, Google, and myriad others.

In a similar vein, it has been contended that the rise of the emerging world economic powers, particularly China, Russia, Brazil, and India, will diminish the importance of the United States in global affairs. This argument is based on the assumption that while the rest of the world continues growing, the United States will be somehow stuck in slow tracks.[28] These scholars underestimate the power of the First Amendment, the openness of American society that not only permits a continuous inflow of hardworking immigrants but also encourages human intellectual capital, scientists, technologists, scholars, and students, to come to America and subsequently build up global brainpower grids with their native countries, thus amplifying the American brain power. America's intellectual network draws brainpower from all over the world, which cannot be said of China, India, or Russia.

A steady, unending stream of dream chasers infuses the aspirational society with a sense of special destiny, of American exceptionalism. Declinists and "patterns of history" determinists also underestimate the self-renewing power of news-media-generated critical processes, as have been seen from crisis to crisis; for example, the Vietnam War, Richard Nixon's Watergate, Bill Clinton's impeachment proceedings, George W. Bush's weapons of mass destruction misinformation, and Abu Ghraib. Each crisis has made the United States more open and stronger and better. The great powers of the past never had the First Amendment as their core value and driving force.

Civic Participation

The free flow of information and openness of culture are central to human freedom and development, but the kind of social sharing and cooperative and voluntary work that people are doing on Wikipedia and other social networks, for example, is not a totally new social phenomenon.[29] Alexis de Tocqueville observed during his visit to the country in the early part of the nineteenth century that voluntary associations, which brought people together for common purposes, enabled Americans to develop the idea of a nation that transcended individualism. This collaborative attitude made possible the development of a civic society with a sense of autonomy, a society that did not have to depend upon the government.

This was also the era of Jacksonian democracy (1824–1854) that favored the common man over propertied elites by expanding the electorate to include all adult white male citizens, encouraged Westward expansion, and developed the idea of America's Manifest Destiny that eventually gave birth to the belief in American exceptionalism, and not least the beginning of the Industrial Revolution in America. Today what some scholars call the social production of knowledge through open source and social network sites is in a sense nothing but the continuation of the old American tradition of sharing and making voluntary contributions to the commonwealth, which the French writer-statesman Tocqueville had astutely observed during his visit to the United States. Wikipedia, an innovative civic participation project, could have developed only in America. Tocqueville wouldn't have been surprised.

The declinists have paid insufficient attention to the core value of American society situated in the First Amendment, the source of energy that incites fearless speech, innovation, and civic participation, which did not exist fully in great powers and empires of the past. If the British Empire were a commonwealth of nations of coequals in power, with the Bill of Rights as its core value, it would have survived. It is in this sense of the First Amendment being at its core value that the United States is historically unique, an historic outlier, and therefore its future cannot be predicted from what happened to other past great societies that thrived on centralized systems of command and control, where critical self-examination through fearless speech in the public domain, and the press, was absent.

We will look at the current field from a different perspective, the perspective of a self-organizing system that is subject to evolutionary growth and constraints from within the system as well as outside forces.[30] The First Amendment–enabled critical examination, which takes many forms, from investigative reporting, whistleblowing, and public opinion polls to politically incorrect comedic speech acts, is internal to the dynamic value system that controls and shapes American society. The culture of innovation in America is enhanced and sustained by the freedom of the press, a most important fundamental right that makes possible unbridled freedom of thought, independent thinking, an open-market-based venture capital system, and a risk-taking milieu that gives everyone a second chance, and most of all it does not stigmatize failures; at the same time it looks askance at the privileges of position and power through public ridicule, satire, and parody.

Most importantly, it is imperative to understand how the constitutional checks-and-balances system keeps American society in a perpetual self-correcting mode, making possible not only the weeding out of the corrupt and incompetent through prolonged political primaries, congressional hearings, and impeachment but also by challenging business and industry in the spirit of marketplace survival-of-the-fittest Darwinism to develop products and services that are safe, consumer friendly, and innovative. If the dynamics of the free speech system has enabled Americans to be continuously innovative and creative—socially, politically, and technologically—the question is whether this momentum can be sustained in the age of globalization, especially when other international actors have begun to play significant roles. How a free society keeps innovating socially, politically, and technologically through the process of public criticism, competitiveness of the marketplace, and cultural openness is a lesson yet to be learned by authoritarian societies waiting for the spring to blossom.

OPENNESS MAKES AMERICANS INVENTIVE

In his introduction to David Brown's exciting book, *Inventing Modern America: From the Microwave to the Mouse*, James Burke, a London-based television producer and author, wrote that the "can-do" spirit and the "readiness to adapt to circumstance is one of America's most enduring characteristics and is what makes the American social environment more amenable to innovation than any other, because America sees change as an opportunity rather than a threat."[31] Henry Ford, for example, adapted a British Royal Navy–developed assembly-line idea of production and used it to usher in an era of "democracy of possessions," which required the application of new technology for making mass consumption products and thus feeding the inventive-entrepreneurial spiral.[32]

Those who do not understand the ethos of American society find it rather puzzling as to what makes an immigrant scientist more productive, inventive, and innovative in the United States than back in his home country. A case in point is Ashok Gadgil, an Indian American scientist at the Lawrence Berkeley National Laboratory in California and one of the thirty-five inventors profiled in David Brown's book, whose invention of a drinking-water purification system showed the potential of saving millions of people in developing countries where diarrhea, cholera, hepatitis, and water-borne diseases are widely prevalent. The "Bengal cholera" of 1992 that spread throughout India and killed approximately ten thousand people challenged Gadgil's sensibilities and inspired him to develop an affordable and effective water purification system using ultraviolet light. Called the UV Waterworks, Gadgil's device today may not be regarded as a disruptive invention, but it is extremely useful for the masses in developing countries where drinking water contamination is commonplace. "You could disinfect one person's drinking supply for a full year for a couple of cents," Gadgil said at his MIT web profile.[33] Perhaps he could have eventually developed this kind of invention in India also, now that the country has begun to gradually open up to the competition of the global marketplace. But India is not an innovation-conscious country. The marketplace is not aggressively competitive. There are hardly venture capitalists in India at present.

If we consider patents as a measure of technological innovation, the United States is the most inventive and innovative nation today. In 2011, for example, 247,713 patents were granted to American inventors, which is 49 percent of the total patents granted in the United States.[34] Burke suggests that scientists and inventors follow the twofold rule propounded by the seventeenth-century French rationalist philosopher and mathematician, Rene Descartes: Be a Doubting Thomas and reduce every system to its fundamentals. You can't sit in your bathtub and wait for *Eureka* to occur. Inventing is sweating all the way (Thomas Edison: "Genius is one percent inspiration, ninety-nine percent perspiration") until a new configuration, a new way of doing things surfaces like an irrepressible force. But you cannot be a Doubting Thomas in a society where minds are closed and markets are controlled and the government keeps a close watch.

With global e-commerce rising, the job marketplace has been changing the world over and demands continuous inventiveness and innovations. A nation that does not invent and innovate will keep borrowing and stealing technology from others. MIT economist Lester Thurow, in the foreword to the book, wrote that scientific and technological advances in most human history have occurred slowly and sporadically in one civilization or the other, often vanishing from the place of its origin, and he warns that there are no guarantees that a nation's technological lead will last forever. But he does not explain why these once creative societies stopped innovating. Consider this, for example: After the decline of Roman civilization, Europe fell under the hegemony of the Catholic Church and gradually slid into what some historians have called the Dark Ages, the period when Europeans' minds were closed and technological leadership passed on for some time to the Islamic world. In some ways China was more advanced in the fifteenth century than in the nineteenth century, but why did China stop innovating? Neither China nor the Islamic world had the culture and institutional infrastructure to accommodate dissent and free inquiry that is a necessary condition for thinking outside the box of religious orthodoxy or the authoritarian canon of absolute rulers.

The restless and questioning spirit in Europe was reawakened with the (re)invention of the moveable-type printing press by Johannes Gutenberg in Germany in the fifteenth century that in the course of time ushered in two great movements, the Reformation and the Renaissance, which caused tremendous intellectual and social upheaval in Europe and consequently the rest of the world and eventually leading to the discovery of America, the colonization of Asia and Africa, and empire building, for good and evil—one might call them early forms of globalization. Liberation from the hegemony of the Catholic Church and the enthronement of reason and empirical methods of discovering knowledge led to an era of discovery that has not come to an end. From the invention of the steam engine to electricity, one great invention of the Industrial Age led to another and propelled some countries to the rank of the first world order and relegated many others to the third world basement.

But what has made scientific and technological innovations a sustainable enterprise in modern times is not only systematic investment in research and development but also academic freedom to pursue inquiry wherever it leads, freedom from the dogma of revealed sources of knowledge. Scientific and technological research is based on an empirical mode of inquiry and experimentation and does not depend upon a priori knowledge or revealed truth, which was the raison d'etre for the establishment of the Royal Society of London for Improving Natural Knowledge in 1662 to "discuss the new philosophy of promoting knowledge of the natural world through observation and experiment."[35]

The motto of the Royal Society, "Take nobody's word for it," is a translation of Latin *Nullius in verba*, attributed to Horace, the Roman poet, who having compared himself to a retired gladiator said that he was under no one's control, that he had no master.[36] The Royal Society's motto was a proclamation that the European society was reclaiming Europe's Greco-Roman culture of free and fearless rational inquiry, which a few years earlier had found a robust expression in John Milton's *Areopagitica* against censorship: "Though all the winds of doctrine were let loose to play on the earth, so Truth be in the field, we do injuriously, by licensing and prohibiting, to misdoubt her strength. Let her and Falsehood grapple; who ever knew Truth put to the worse, in a free and open encounter? . . . Give me the liberty to know, to utter, and to argue freely according to conscience, above all liberties."

This was the beginning of the end of Europe's long, drawn-out Dark Ages and the start of the Age of Enlightenment. Once again Europeans opened their minds to new ideas, and since then they have not stopped innovating and inventing. The Enlightenment was nothing but emancipation from blinding religious faith, deliverance that eventually led to the foundation of the First Amendment. Today research universities have become a most prolific source of inventiveness and innovations in the United States, which attract the best brains from all over the world, thus making the United States the most inventive and innovative society today. In 2011, universities in the United States reported to have earned $1.8 billion in royalties from their inventions.[37] This happens not only because of the abundant research funding resources but also because the First Amendment and American do-it-yourself culture creates a feeling of human dignity and expansiveness. That explains why Gadgil was celebrated as an American inventor because the United States provides the socially supported scientific platform built on the pillars of the First Freedoms. We know, for example, how after World War II German rocket scientists were persuaded to come to the United States to help beat the Soviet Union in the race for space, and even today research doors remain open to outsiders.[38]

Of course, what has been gained can be lost, but only if Americans close their doors and their minds and lose their competitive spirit. From ballparks to technology parks, competition, the Olympian spirit, is at the heart of American society, which cannot flourish without the accountability of the free press. Britain and Germany were the technological leaders since the Industrial Revolution began two hundred years ago, and they still continue to be some of the top inventive societies. What makes the United States so different from the past technological leaders is that it is culturally situated between First Amendment freedoms and the awesome temptations of the open competitive marketplace, the genesis of the aspirational society.

The free marketplace of goods and ideas has been fuelling the relentless pace of inventiveness and innovations in the United States and has created tremendous economic prosperity. Nevertheless, in order to sustain its innovative power the United States needs to expand its collective brainpower beyond its borders, as corporate America has been doing through insourcing of talents and outsourcing of jobs, a two-way traffic, augmented by emerging technologies such as grid computing. Job outsourcing and grid computing are parallel concepts, and they are the objective correlative of a potluck dinner party—pooling divergent expertise and scaling up resources for cost sharing and problem solving with speedy delivery—a global projection of Tocqueville's civic society in America.

GLOBAL BRAIN-GRID EMERGING

A few years ago, IBM, Intel, Hewlett-Packard, and Sun Microsystems announced the formation of a consortium to sell to the corporate world the idea of adopting open-source grid computing for commercial applications.[39] The initial barrier to the widespread use of the concept of the grid might have been psychological, implying that *grid* means "gridlock." But it's an innovative idea whose time has come, and it is very simple. When you turn off your computer and go home, no one can use the computing power locked in on your office desktop. Nor can the dispersed unused computing power of hundreds of other computers be used or leased by your company to anyone, which is such a waste. If you multiply this unused computing power by millions of computers around the world, you get an idea about the extent of unused resources. This is a rather too simplistic explanation of grid computing; nevertheless, it gives us an idea as to what distributed computing, like distributed work or job outsourcing, can do for an organization.[40]

Grid computing is the application of pooled computing power to a single technical and scientific problem that requires working with multiple databases. Grid computing pools the power of geographically dispersed computers and divides and distributes a single, multidimensional problem to computers on the grid for simultaneous and parallel processing. Grid computing is a kind of controlled, networked collaborative computing that focuses on a complex problem, which might require working with hundreds of dispersed databases simultaneously. Heterogeneity and geographical dispersion of computers on the grid technically distinguish it from cluster computing, but more importantly it is the development of collective brainpower to solve problems on a global scale. Grid computing might include the resources of supercomputers, cluster computing, and cloud computing (which is nothing but a kind of storage system, data warehouse, software and apps platform), but the purpose is to coordinate and multiply brainpower, both computer and human, to solve complicated problems in the shortest possible time. Never was the advice of Benjamin Franklin truer than today when in *Advice to a Young Tradesman, Written by an Old One* (1748), he said, "Time is money."[41]

In August 2001, the National Science Foundation undertook an ambitious project, TeraGrid, by virtualizing geographically dispersed supercomputing centers into a grid, which made it possible to make them work as if they were one giant virtual computer. Virtualization is an Internet technology approach that optimizes, pools, and shares resources so that supply and demand match each other as needs change. TeraGrid, replaced by Extreme Science and Engineering Discovery Environment (XSEDE) in 2012, was capable of creating several times the power of the most powerful supercomputer, proficient to carry out trillions of operations per second. Grid computing enables not only geographically dispersed computers but also different operating systems to coalesce as one brain and provide the user a steady, consistent, and economical access to enhanced resources from any point of access on the grid. The unleashed supercomputing power, emanating from a single unified source, could be used for solving large-scale, data-intensive science applications, such as designing new drug remedies based on molecular modeling, tsunami-earthquake modeling and forecast, and military operations in insurgency-ridden areas. In a grid-based virtual laboratory, a scientist could examine millions of molecules in the Chemical Data Bank to identify and select those that have the most potential use for designing a new drug for a specific type of cancer for each individual cancer patient, for example. Grid computing may be the answer to the challenge of ever-growing Big Data, the black hole of the digital age, in a manner of speaking.

The grid power could be used for teleimmersion, giving users high-bandwidth access to a virtual environment such as a simulated surgical operation or a rising pandemic like swine flu. Eventually all computers will be linked with local, national, and international grids, enhancing computing power and sharing databases and applications that may reside anywhere. Just as TCP/IP protocol broke barriers and made the Internet possible, grid computing must have a single, worldwide standard, and most bets are on the Globus, an open source system that would make it possible to tap the Internet on demand from anywhere.

The Globus Toolkit, a collection of software applications and resources to support grid computing, could locate where a particular database is housed; how to divide and distribute given computing work among several computers on the grid system; and whether the user is an authorized person. Once a handheld or desktop machine is plugged into the Globus, it could draw not only its computational power from the grid but could also use myriad applications, such as audio-video streams, databases, videoconferencing, and so on. A small company needing a petabyte of computing power could simply tap into the grid to perform a specific task and once the job is done, the resource is returned to the grid for others to use. Universities could form their own grids and be networked into national and international grids, creating the possibility of on-demand grid brainpower. The central processing unit need not be on every desk, but those who have it could share it with others. The cell phone will become a gateway to a supercomputing brainpower grid.

The wireless revolution that is sweeping the world now will make grid computing a global phenomenon. According Internet expert Larry Smarr, "Because of the miniaturization of components, we will have billions of endpoints that are sensors, actuators and embedded processors. They'll be in everything, monitoring stress in bridges, monitoring the environment, ultimately, they will be in our bodies, monitoring our hearts."[42]

The foundation for grid-computing infrastructure must be based on security as the first principle, of course. Some regard grid computing as the natural evolution of the Internet-networked world. It is one of those technologies whose value is perceived to be so great as to make its universal acceptance inevitable. When all major research universities and technology institutes are virtualized into one international computer grid, the impact of the emergent brainpower would be immeasurable. Think about what it would do to the business world when it goes on the grid. Imagine global retailers like Walmart going on the grid to make their supply-chain systems more efficient and greener. Grid technology might enable small countries like Bangladesh and Vietnam to compete with Chinese manufacturers and other global chain-supply juggernauts. Size and geography won't matter in the age of emerging global brainpower and distributed work, where the United States' open, free market culture will, however, play a pivotal role and keep attracting the best brains from all over the world. Geography need not determine a country's destiny in the digital age.[43]

DEMOCRATIZATION AND INNOVATION

The research of European scholar Mario Coccia of the National Research Council (Italy) suggests a strong correlation between democratization and technical innovation. According to him, "Most free countries, measured with liberal, participatory, and constitutional democracy index, have higher technological innovation than less free and more autocratic countries, so that the former have a higher interaction among social, economic and innovation systems with fruitful effects on economic growth and the wealth of nations."[44] The three indices that Coccia uses to measure the democratic quality of a country include the Freedom House Index of liberal democracy; the Polity Index IV of constitutional democracy of the University of Maryland and Colorado State University; and the Vanhnen Index of participatory democracy of Helsinki University and the International Peace Research Institute, Oslo. Although these indices are based on a different set of questions, they measure the same underlying democratic quality of a society.

Coccia measures technological innovation primarily in terms of the number of patents and the payments of royalty and license fees earned by a country, supplemented by other indicators such as scientific and technical journal articles and researchers and technicians employed in research and development and R&D outlays. Based on his empirical research Coccia concludes that on the whole free countries as measured by the three liberal, participatory, and constitutional democracy indicators have a higher positive correlation with technical innovation than the less free ones.[45]

There is plausible historical evidence that suggests that the liberalization of Western society beginning with the parliamentary constitutional monarchy and the Bill of Rights in England in 1689, the French Revolution (1789–1799), and the ratification of the U.S. Constitution along with the Bill of Rights (1791) that guaranteed First Amendment freedoms among other rights, preceded and overlapped the Industrial Revolution in the eighteenth and nineteenth centuries. The slow and gradual liberalization of Western society runs on a parallel track with the growth of industrialization. The essence of liberalization and democratization is freedom of the press without which diffusion of knowledge about innovation and benefits to people and consequently the enlargement of the marketplace cannot occur. One can see how liberalization, freedom of expression and other individual rights, entrepreneurship-based technological innovations, and the marketplace have been feeding on each other to effect a dynamic system of growth that is still continuing. The role of free speech in the dynamic loop constituting entrepreneurial innovation and the marketplace is to provide feedback and corrective mechanism for growth and sustainability. It is a question of control and freedom, according to Lawrence Lessig, who says, "If communism vs. capitalism was the struggle of the 20th century, then control vs. freedom will be the debate of the 21st century. If our question then was how best to control, our question now will become whether to control. What would a free resource give us that controlled resources don't? What is the value in avoiding systems of control?"[46]

Of course, there is more to innovation than freedom and control. It is the tolerance of failure of which Silicon Valley is an exemplar, where failure is not only not stigmatized or punished but in fact it's a necessary condition for success. Craig Newmark, the creator of Craigslist, wrote in his blog that in 2009, when he attended the Reboot Britain conference, he

> was struck by the repeated comment that failure is stigmatized in UK business culture. In Silicon Valley, failure is just a normal phase of one's career. You might succeed in your first endeavor, probably not, so you're ready to persist in subsequent efforts. That is, there's some expectation of failure and the expectation that you'll get over it. This is not unique to Silicon Valley, but it's far more expected here than anywhere I've heard. The attitude is the norm here, but in a lot of places failure continues to be stigmatized, and it's hard to recover. It seems that widespread innovation and success requires the acceptance of failure, and then a readiness to move on. That's generally true in Silicon Valley, maybe needs to be true in the UK and maybe everywhere else.[47]

The First Amendment has turned America into an open trial-and-error experiential universe, where you learn at your own risk. One of the most important and crucial functions of the free press in the United States is to test the veracity of claims whether it is a corporation, inventor, scientist, discoverer, or politician who makes them. This kind of adversarial relations, trial in the media, is further reinforced by the American adversarial judicial system, as we saw in the case of *JB v. MB*. Claims about grid computing, space exploration, genetic tweaking, stem cells, sperm-and-eggs, or the nature of motherhood are all contestable in the media and the court. This is not the portrait of a harmonious society; rather, it's a society undergoing a perpetual labor of creativity. In such a society, success is not an end state in itself; rather, it is a halfway house for something better to come. The openness of the United States puts it in a perpetual beta mode. The final product has yet to come if the sources of information and the news media remain, as we discuss in the next chapter, "unfettered, robust, and wide open."

Chapter Three

Telling Stories to a Hyperconnected World

A VIGOROUS AND WIDE OPEN NEWSOSPHERE

What if Gordon Gekko in *Wall Street* and Richard Nixon in the White House were to collaborate? Would American democracy survive? Alexis de Tocqueville could not have foreseen the danger. But Justice Potter Stewart did see the danger, and he warned us in the *Pentagon Papers* case in 1971:

> In the absence of the governmental checks and balances present in other areas of our national life, the only effective restraint upon executive policy and power in the areas of national defense and international affairs may lie in an enlightened citizenry—in an informed and critical public opinion which alone can here protect the values of democratic government. For this reason, it is perhaps here that a press that is alert, aware, and free most vitally serves the basic purpose of the First Amendment. For without an informed and free press there cannot be an enlightened people.[1]

Now the danger to the press is existential and comes not from the government or Wall Street but from the burgeoning new communications technologies. With grave apprehensions, journalists, media scholars, and news media establishment have been discussing the slow and steady decline of the traditional news media due to the challenge of the emergence of online integrated news media, wondering what might happen to the world if the hard-copy, print version of newspapers and magazines and regularly scheduled news broadcasts dwindled or disappeared eventually.[2] Will the press continue to be "uninhibited, robust, and wide open" for democratic deliberations, serve as a check on the unbridled corporate and political power concentrated in the hands of a few, and, most importantly, stimulate the innovation necessary for the aspirational society?[3]

In response to the existential challenge, daily newspapers have been innovating into multimedia, consolidating their operations and in some cases shuttering down. Some like the *Christian Science Monitor* have gone completely online, while others have adopted a dual mode of delivery, a few days of hard-copy delivery and the rest online. New Orleans is the first major metro area in the nation that will be without a daily newspaper since the 175-year-old newspaper *The Times-Picayune* decided to go online in fall 2012, limiting its print edition to just three days a week. How much will the "civic common" shrink, and what will be its effect upon the social life of the city, has yet to be seen.[4] *Newsweek*, once upon a time a global journal, made a brave attempt at repurposing and rebranding to consolidate and hold on to its shrinking base of readership, but it gave up and has gone into the belly of the online *Daily Beast*. In search of niche market power, broadcasting media have been proliferating and narrowcasting to find an audience "fit though few" who can sustain them financially.[5]

Journalism is certainly not dying, but if some of its crucial newsroom functions, for example, keeping a hawk eye on public affairs, seeking leaks, and raking muck to expose the misuse of power, are compromised due to cutbacks, it might become just another business, selling itself to the highest bidder.[6] The business of chronicling the world honestly and accurately is a serious business, and we need both professional as well as amateur journalists to trawl the globe and make it knowable, transparent, and governable according to the free will of the people. Media scholars Leonard Downie Jr., and Michael Schudson, in their report on the state of American journalism, emphasize the indispensability of robust investigative news, "the watchdog function of the press—reporting that is aggressive and reliable enough to instill fear of public embarrassment, loss of employment, economic sanctions, or even criminal prosecution in those with political and economic power. Watchdog or accountability reporting aims to foil the arrogance of power and self-dealing rather than to advance ideology or policies. It holds government officials accountable to the legal and moral standards of public service and keeps business and professional leaders accountable to society's expectations of integrity and fairness."[7]

People committed to the competition of the news marketplace believe that there is an urgent need for developing fresh approaches and building new business models to turn emerging opportunities into profitable businesses. A free press cannot survive on public funding alone, in spite of the laudable success of PBS, BBC, and new ventures such as ProPublica. Unfettered journalism in order to serve the public interest can thrive only in a Darwinian marketplace, which means that news organizations must compete for audiences by offering the best in quality and breadth of their coverage, including the explanation of complex issues, investigation, analysis, and local knowledge.

The emerging technology of online journalism is transforming news gathering and dissemination and how people think about and seek information and communicate and interact with each other. The instant globalization of news is on the rise if the events of the Arab uprising, bloody and exhilarating, are indicative of future trends. It has become vitally important that we focus on how online journalism, including civic media, blogs, Twitter, and myriad social networking sites powered by mobile technologies, is challenging and competing against the traditional news media in its legacy role, especially regarding investigative reporting, as public eavesdroppers and auditors in an open democratic society. The dialectics of challenge and response between the traditional and new online media, in the emerging dynamics of the news and information ecosystem, might give rise to hybridization, co-optation, or synthesis of the old and the new.[8]

News media organizations that produce newspapers, news magazines, and broadcasting news are mostly interested in hard news, which requires time-consuming efforts, footwork, telephone calls, cultivating sources, interviews, and database research in digging out facts and doing investigation; in the process they also gather soft news, celebrity profiles, and gossip that draw into their folds leisurely readers. At present it seems that it is mostly the soft news that is going away to the Internet along with some advertising revenues and valuable consumer databases. Search portals and social networking sites, Facebook, Google, Yahoo!, Craigslist, and innumerable others are nibbling at the advertising-based business model of the traditional news media, with tremendous impact on the newsroom. Traditional news media organizations are not only indispensable for a healthy democratic society but also for a robust marketplace where innovations, disruptive and incremental, are tested. Traditional news media must innovate for making profits as well as for serving the public interest.

Unfiltered and Unbridled

While traditional American news media organizations are struggling with their future as viable businesses, more importantly, the global reach and influence of the new generation of cyberjournalists has seized the attention of governments around the world, particularly authoritarian and repressive regimes who have instituted countermeasures for controlling the flow of unfiltered and unbridled news.

The Committee to Protect Journalists reported that 179 journalists worldwide were put in jail in 2011. Iran with forty-two led the way in jailing journalists, followed by Eritrea, China, Myanmar, and others. In 2008, for example, of the total imprisoned journalists, fifty-six were online journalists, a growing tribe of citizen journalists that authoritarian governments find it difficult to ignore.[9] Authoritarian as well as democratic governments have to face the ever-expanding freedom of cyberspace made possible by diminishing barriers to access the Internet and the availability of inexpensive and mobile tools such as cell phone cameras, Wi-Fi, GPS, and encryption to anyone who wants to blow the whistle for his conscience and the country. Rootless and footloose freedom is not only strengthening the traditional news media in many ways by providing them with raw footage of unfolding events, as it happened in the Iranian postelection protests in June 2009 and the uprisings in China's far out-of-sight western Xinxiang Muslim province in July 2009, and frequent uprisings and self-immolations protests by Tibetans everywhere, but it is also giving rise to millions of grassroots civic journalists all over the world, who believe that they must tell their stories to the networked world.

The unstructured narrative freedom to show and tell a breaking story any which way, conventional and unconventional, and the technological means of gathering and disseminating unheard silences and suppressed voices instantly and globally on multiple platforms is ceaseless and breathtaking. But it is beyond the reach of the traditional news media to do so unless it co-opts and incorporates them into working relations and a new business model. That is the challenge. Today cyberjournalism makes the lives of the people in power extremely uncomfortable when stories are published instantly that would have normally passed through a well-established editorial supply-chain system, occasionally massaged by corporate and state authorities. On the other hand, since cyberjournalists lack the financial and legal protection of the news establishment and also, albeit indirectly, of the government, they face greater harassment than traditional journalists.

Harassment

Even in a most open society built on the foundation of the First Amendment freedoms, as is the United States, journalists are sometimes harassed or flattered and co-opted by authorities, though in rather more urbane ways. They are, for example, subpoenaed and questioned about confidential sources used in their reporting, sources that are cultivated to uncover corruption, and in some cases in which they are known to be more knowledgeable about a crime than professional crime investigators.[10] Good journalists report not only about open and observable events, traffic accidents, crimes, and what the First Family dog ate for breakfast, they must also investigate and report the information that impacts society, the information that must be vetted. Not to report the truth about corruption, for example, would not only be complicity in crime but also an extremely unethical and unprofessional behavior for journalists. When a journalist suspects something wrong in the public sphere, he must investigate, whatever it takes, as David E. Sanger of the *New York Times* did in his book, *Confront and Conceal*, detailing President Barack Obama's foreign policy, including the use of drones against enemy combatants abroad and cyberwarfare against Iran's nuclear facilities.[11] His book, based on deep reporting rather than on whistleblowers' leaks, left him unharmed, though not everyone is so lucky.

Jim Taricani, a Rhode Island television reporter, was put under house arrest for six months in 2004 for his refusal to disclose the source of a videotape showing a state official taking bribes from an undercover law enforcement informant.[12] Judith Miller, a Pulitzer Prize–winning journalist who worked for the *New York Times* until 2005, spent eighty-five days in jail in 2005 for refusing to testify before the Grand Jury investigating a leak about CIA agent Valerie Plame Wilson, though subsequently there were serious questions about her integrity as a journalist.[13]

Whistleblowers leak documents anonymously or talk to reporters on the promise of confidentiality, though some of them have ulterior motives and use journalists to advance their interests and vice versa. Online journalists, news bloggers, Twitterers, and social media activists are under no obligation to do any investigative reporting, though some of them do; most do not have the training and the financial resources to do so. Traditional journalists, on the other hand, are trained to believe that accepting a hazardous assignment for the sake of the public good is not only professionally necessary but also highly moral, though underneath the veneer of altruism there might lurk ambitions of personal fortune and careerism. Those who defy authorities and serve the cause of freedom of the press do grab media attention and become celebrity journalists, and they earn national recognition and huge financial rewards through book contracts. What would have been the future of Bob Woodward and Carl Bernstein as journalists without the legend of Deep Throat?[14]

Sometimes journalists clandestinely collaborate with the authorities and reap great rewards through access to privileged sources and earn media accolades, and they become quotable sources themselves. Judith Miller, however, could not hide behind her confidential sources and became a mortifying source of embarrassment for the *New York Times* and her admirers. A blogger or an independent journalist could not have gotten away with her high-powered tales regarding weapons of mass destruction that sounded to be well researched and thoroughly vetted. As Franklin Foer wrote in *New York Magazine*, "Pulitzer Prize winner Judith Miller's series of exclusives about weapons of mass destruction in Iraq—courtesy of the Iraqi Shia politician Ahmad Chalabi—helped the *New York Times* keep up with the competition and the Bush administration bolster the case for war. How the very same talents that caused her to get the story also caused her to get it wrong."[15]

A Hazardous Vocation

For decades the courts in the United States have been wrestling with the issue as to how to balance competing fundamental rights; for example, protecting the news media's obligations to do reporting by discovering and cultivating confidential and reliable sources and the equally compelling need of the judicial system for accessing critical information that might not be available anywhere else, which only a reporter might possess. The fair trial requirement of the Sixth Amendment is most compelling: "In all criminal prosecutions, the accused shall enjoy the right to a speedy and public trial, by an impartial jury of the State and district wherein the crime shall have been committed, which district shall have been previously ascertained by law, and to be informed of the nature and cause of the accusation; to be confronted with the witnesses against him; to have compulsory process for obtaining witnesses in his favor, and to have the Assistance of Counsel for his defence."

In certain circumstances courts feel compelled to demand information from news media organizations including outtakes and unpublished materials to uncover evidence that might impact the case under trial. The courts issue subpoenas to get the crucial information in the possession of a journalist without which justice might be miscarried. And in the case of noncompliance, they exercise the contempt power. Many First Amendment scholars and news establishment supporters argue that subpoena and contempt power used against journalists endanger freedom of the press because journalists might become too cautious in cultivating confidential sources. Without going behind the scenes and investigating stories, they might have to feed on public relations handouts instead of finding out the truth about the misbehavior of prominent public and private figures as well as corporations. Every year hundreds of American journalists receive subpoenas, according to Agents of Discovery of the Reporters Committee for Freedom of the Press. State shield laws enacted to protect journalists from unnecessary disclosures of confidential sources do not go far enough. "In opinion after opinion," said the report, "judges fail to acknowledge any special role for the media in a democratic society, or any public interest in ensuring that the media remain impartial and disinterested both in perception and reality."[16]

Balancing Competing Interests

But the promise of absolute confidentiality to sources does create legal and moral hazards. What if a reporter is the only source of information that constitutes a crucial piece of evidence in a legal case, information so necessary that it is indispensable to the case, and unless the source is questioned there is the danger of justice being denied? In such special circumstances the source, of course, must be revealed regardless of the promise of confidentiality by the reporter simply because the right to a fair trial is no less important than freedom of the press.

In the course of time, however, the U.S. Supreme Court developed a "preferred position balance theory" in deciding conflicts between freedom of the press and other fundamental rights.[17] In numerous decisions, the U.S. Supreme Court held that the First Amendment freedoms are the foundation of a free society and consequently deserve more enhanced protections than other constitutional rights. Freedom of the press, nonetheless, does not diminish all other fundamental rights, not least the constitutionally guaranteed right of a person to a fair trial that entitles the accused to have access to all crucial evidence, including in the possession of a journalist. By according freedom of the press a preferred position in balance with other fundamental rights, the Supreme Court nonetheless obligates the government to bear the burden that forcing a journalist to disclose his confidential news source is absolutely necessary in a specific case. Freedom of the press is protected, though not absolutely.[18]

Much has been said about how Bob Woodward and Carl Bernstein of the *Washington Post* depended upon their well-managed, confidential source, Deep Throat, for their historic investigative reporting about the Watergate affairs that brought President Richard Nixon down and reset the nation on the path to political innovation and self-renewal.[19] In 2005 Deep Throat, the FBI's deputy director William Mark Felt, outed himself as the person whose Watergate confidential revelations caused so much national soul searching and consequently strengthened democracy by planting seeds of doubt in the American collective consciousness that the government can do wrong and will do wrong. Corporate whistleblowers disclose corrupt accounting practices, faulty products, and other malfeasances to journalists so that society might benefit, for example, as happened in the case of Enron, the Houston, Texas–based energy company that used fraudulent accounting methods to establish its position as a global market leader. The 2001 Enron scandal (along with Tyco International, Adelphia, WorldCom, and others) was one of the greatest public outrages that shook up America and led to the creation of the Sarbanes-Oxley Act of 2002 that requires greater transparency from U.S. public corporations, including their operations in foreign countries.[20]

News media, nonetheless, must not act as unofficial investigators or information gatherers for law enforcement. When the government and corporations invade newsrooms and tamper with news-gathering processes, freedom of the press is compromised. If the information obtained from confidential sources is revealed without any compelling reason, it damages the reporter-source relationship and tarnishes news organizations' image of independence. Trust dissipates when confidential sources suspect collusion between law enforcement and news organizations. Future access to reliable information is blocked, increasing the possibility that people in power might get away with corruption. Likewise, when readers suspect that advertisers are influencing the publication of news and commentaries, they lose trust in the news media.

When we applaud the rise of cyberjournalists, the question is whether they can establish trust at a time when traditional news media are failing in keeping up the public trust. Commenting on the report about the former CBS anchorman Dan Rather's faulty reporting—that unfortunately ended his long, glorious career as a journalist—about George W. Bush's Guard duty during his days in the Texas Air National Guard, Jay Rosen, journalism professor at New York University, said, "I hope that broken contraption 'trust us, we're CBS,' forces the network into the clear skies of a new idea: We used to do our reporting in a way that required the public to trust us, their professional journalists. It worked for a while, but times change. Now we have to do our reporting in a way that *persuades* the public to trust us. CBS News: are you up to it?"[21]

Dan Rather was undone by bloggers. Harry MacDougald and other bloggers posting on the Free Republic website questioned the authenticity of the so-called Killian documents, the basis of the CBS report that claimed that the information came from a reliable source, the personal notes of the late Lieutenant Colonel Jerry B. Killian, George W. Bush's squadron commander during his Air National Guard service.[22] CBS and its source could not provide the original document, the forgery allegation went viral, and the grand old man of the news media Dan Rather saw the terra firma of journalism slipping underneath his feet, reducing him to wails and plaints of a man "more sinn'd against than sinning," seeking redress in litigation against the very media company that had given him a voice of authority for decades.[23] In a densely networked world, it is difficult to be heroic for too long. But the aspirational society needs no heroes.

Trust, But Verify

Cyberjournalism is no doubt changing the climate of news gathering and truth telling. Online journalists who may originate their investigating reporting from anywhere in cyberspace can provide readers with unfiltered news and challenge the traditional news media to launch their own investigation. It was the online *Drudge Report*, for example, which was the first news source in 1998 to break the Monica Lewinsky-Bill Clinton clandestine liaison to the public after *Newsweek* had declined to publish the story.[24] In the aspirational society, uncovering the truth about the seekers of power, money, and influence matters most, so that the society does not regress into a recurrence of the rise of men like Enoch Lewis "Nucky" Johnson of *Boardwalk Empire*.[25] Free speech can break the patterns of history.

But one has to be careful. Occasionally, online journalists seem to rush in where traditional journalists hesitate to go, sometimes transforming unfiltered news into a digital theater of grand and cruel gestures. A well-rehearsed and structured tableau of victims and perpetrators locked in murderous action, for example, could be uploaded for instant, global, synchronous reactions of curses and condemnations, revenge and retributions, blood for blood, as we saw ethnic Chinese Han settlers demanding vengeance against the Uighurs in 2009 in Xinxiang.[26] Anyone who wants to make a political impact without the mediation of journalists could develop a digital strategy to be on the world stage.

The militants in Iraq who beheaded the twenty-six-year-old American contractor, Nicholas Berg, in 2004 did not need a journalist to tell their story.[27] They videotaped the gruesome killing and let loose the images in cyberspace as a vindictive response to the barbaric acts of a few U.S. soldiers at the Abu Ghraib prison. Although U.S. television media withheld from the public how Berg's head might have been cut off, sooner or later Americans would have discovered the cybercasting of terror. And millions did, as if it were another grim reality show. Americans were mutilated, dragged, and hanged for the world to see how the forces of liberation and democracy were performing in the streets of Baghdad. The Arab-Muslim world's reaction seemed like subdued approval. Most Americans did not know what was happening at Abu Ghraib and were shocked at the revelations. Shock and awe, the military doctrine that has become associated with former defense secretary Donald Rumsfeld, is the new emerging facet of the online news ecosphere.

In times like these when we are deluged with an unceasing barrage of commentaries from empty talking heads and garrulous pundits, one might say that digitally captured significant moments can tell the real story with telling effects. Whether you hear the click or not, somewhere a miniature digital camera might have taken a snapshot of some fleeting, unguarded moment that might become a witness to an atrocity or crime, or the self-disclosure of the inner thoughts of a politician seeking a public office. But before the image or video clip finds its way to a newsroom, inquiry commission, or court, the public would have the chance to see it in cyberspace, where the state classification system or court gag orders don't work. It's journalism without journalists, which nonetheless cannot always guarantee the veracity of its stories and establish trust, especially when it shows digital humanity "red in tooth and claw."

Consider this, for example: Shortly after the picture of the abuse of Iraqi prisoners by the U.S. troops were published in 2004, the *Daily Mirror*, a London tabloid, published photos of Iraqi prisoners being abused by British soldiers of the Queen's Lancashire Regiment, which subsequently turned out to be fake. Instead of verifying the veracity of the pictures in the light of mounting expert criticism, as a responsible editor should have done, Piers Morgan, the *Daily Mirror's* editor, insisted that the pictures were genuine and the newspaper's account of the abuse was accurate.[28] Though Morgan lost his job for publishing the unverified story, the warning cannot go unheeded that in the Internet age, the responsibilities of media gatekeepers, anchorpersons, and editors have become all the more crucial. Otherwise, a newspaper is no different from a wayside blog. The camera does blink and lie.[29] Sources do tell lies. Making digital distortion and manipulation is easy. By the time the *Daily Mirror* realized its error the damage had been done. Apologies cannot erase networked digital memories because databases are indestructible, almost. Cut and paste can be done with a few clicks and uploaded for the world to see.

On the other hand, consider this: if camera cell phones were available during the Nazi concentration camps, the Holocaust might have been a much more horrendous story. The Vietnam War is engraved in two images: a little Vietnamese girl running scared and stark naked on a highway after a napalm bomb attack; and the other, a Viet Cong prisoner being shot point blank.[30] But today such atrocities can be uploaded to the Internet without the mediation of photojournalists. Regarding the infallibility of the mainstream news media, in 2005 *Newsweek* reported a Koran desecration story in Guantanamo Bay prison that led to deadly anti-American protests involving at least seventeen deaths in Pakistan and Afghanistan. The desecration was based on anonymous sources, which turned out to be false. *Newsweek* made a feeble attempt to apologize, but the attitude that "we the media" can do no wrong has persisted.[31]

Outrageously Transparent

On January 24, 2004, U.S. Central Command disclosed to the news media its ongoing investigation regarding the torture and abuse of Iraqi detainees at Abu Ghraib prison, which surprisingly caused little outrage in the media until CBS *60 Minutes II* broadcast the gory photographs of torture and sexual abuse. If instead of a digital camera, which has no negative and needs no development, an analog camera had been used at Abu Ghraib prison where Iraqi detainees were subjected to extreme cruelty and inhuman, degrading sexual abuses including rape, sodomy, and homicide by U.S. military police and other U.S. agency personnel, someone probably might have destroyed the negatives and stopped the picture development for fear of exposure. Even deep journalistic accounts or a book wouldn't have captured the sexual humiliation and mortification of Iraqi prisoners whose only fault was that they were captured and wouldn't respond to military intelligence experts' extreme interrogations.[32] Regardless of the Geneva Conventions, torture of prisoners of war has been an ignored but universal practice, but now that unobtrusive digital cameras can not only capture the abuse but also instantly put it into cyberspace, thanks to smart phones and other mobile devices within the reach of most people all over the world, transparency and instantaneity have become a modern condition of living and a tremendous challenge to traditional journalists.

Some good did come out of the evil deed: the Bush administration secretary of defense Donald Rumsfeld rushed to Baghdad and ordered hundreds of Iraqi prisoners to be released. In different ways, as shock and awe gradually began to sink into the American collective psyche, Americans expressed their extreme outrage at what had happened at the Abu Ghraib prison.

After the prison abuse revelation, late Senator Ted Kennedy took to the floor of the U.S. Senate on May 10, 2004, and said: "On March 19, 2004, President Bush asked, 'Who would prefer that Saddam's torture chambers still be open? Shamefully, we now learn that Saddam's torture chambers reopened under new management: U.S. management.'"[33] The liberators became torturers instead of propagators of freedom and democracy in the Arab-Muslim world. Some might say that perhaps this was an uncalled for and excessively vituperative overreaction by a liberal Democrat, but what better way to tell that the aspirational society was losing its moral compass than to indulge in theatrical hyperbole and a grand gesticulation of outrage!

The other form of outrage came from Republican diehards who were shocked that someone allowed those sadomasochistic pictures of prisoners to be shot at all, and worst still, why no one ever stopped them from finding their way into the news media. They failed to understand that in the digital age it is impossible to block information or images from being shared. The Internet is a shareable, collaborative, and more or less uncensorable medium. Whatever can be digitized will find its way on the Internet, where shared transparency is the norm, and it is a serious challenge to traditional journalism regarding its trust building and gatekeeping functions.

CONTINUUM OF THE FAKE AND THE REAL

How should a person distinguish between the fake and the real in cyberspace? In the age of the Internet, although one does not need to be an established journalist to get published or broadcast, the question of building trust is of paramount importance. In the world of 24/7 instant news, as we saw in the case of the *Daily Mirror* and CBS's Bush National Guard story, journalists sometimes give up their gatekeeping functions too hastily. And in the process of making an exclusive story, they abuse people's trust, with far-reaching consequences. It's not only about the Bush administration's well-orchestrated pre–Iraq War propaganda about the weapons of mass destruction (WMD), which most U.S. media accepted without questioning, and later on began to discover to their dismay how wrong they were. Some of them, ironically, the *New York Times* and the *Washington Post*, for example, humored themselves in the luxury of critical self-examination, and they published lengthy analyses about why they ceased to think independently and succumbed to the official line about WMDs. The post-9/11 U.S. media mind-set turned out to be tragically dangerous to the world.[34]

The behavior of U.S. media especially after 9/11 demonstrated that freedom of the press does not necessarily lead to independent and objective reporting and analysis. U.S. news media organizations, especially wire services such as the Associated Press, dominate worldwide news gathering and dissemination, and one might call them the eyes and ears of the world, in the sense that we see the world the way they want us to see it.

Unreliable Sources

But if news organizations feed on official handouts as reliable sources, how can you always trust them? It is not about their biases and prejudices in the selection and exclusion of stories, which is not an insignificant problem; rather, it is about the mere act of reporting facts that must be verified instead of relying upon "reliable sources" that cannot be named because of the promise of confidentiality. When the entire news media collectively develops a closed mind-set and dances the same dance and talks the same talk, the dictum "Trust, but verify" does not apply because the incentive and intermedia competition to get the story first becomes more important than to get it right.

In 2004, based on web sources, U.S. news media reported two beheadings by Islamic terrorists in Baghdad. The first one turned out to be a Web hoax and alarmed the media about the haste in publishing stories without any verification or caveat. And a week later, when another Islamic website reported the beheading of a CIA agent, the Associated Press did not know what to do about it; nonetheless, it published the story as something newsworthy, hedging that the authenticity of the tape could not be verified immediately, and at the same time the wire service quoted a CIA official's denial that any of its personnel had been missing.[35] Since Islamic terrorists have used the Internet to post videotapes of beheadings as well as their jihadi announcements, obviously the wire service, it seemed, had no choice but to report the latest web event, which included masked men surrounding the victim, with one of them holding a knife against the latter's throat. But the Associated Press had a moral responsibility to go behind the story and uncover the truth.

Only a week before the story of the CIA agent beheading, the Associated Press, Reuters, and other media outlets had reported the beheading of Benjamin Vanderford, a twenty-two-year-old man from San Francisco. But that was actually based on a video clip that Vanderford and his friends had made and placed on Kazaa and Soulseek file sharing websites. As the *San Francisco Chronicle* reported, the pranksters just wanted to see how fast the story was picked up on by the world media.[36] It seemed that reporters, instead of doing journalism the old-fashioned way, deep reporting based on legwork, interviewing multiple sources, and making phone calls, simply surfed the web for news gathering. That's not journalism.

Since 9/11 Western media have been paying serious attention to Islamic websites for any lead that might be useful in news gathering about Islamic terrorists instead of doing their own independent investigations. Ironically, when an Arab Islamic website, the Al Minbar, published the fake Vanderford beheading story, the mainstream news organization could not have ignored it for fear of being left out of a big story. In light of earlier beheadings, for example, that of a *Wall Street Journal* reporter Daniel Pearl in Pakistan in 2002, and three other hostages including American businessman Nicholas Berg, South Korean translator Kim Sun-Il, and Bulgarian truck driver Georgi Lazov, reporting the story about the latest beheading was based on its plausibility rather than its verifiable authenticity, which is poor journalism. Of course, it is a major dilemma for news media about how to deal with events that happen at lightning speed, but at the same time, it should be kept in mind that disclaimers and caveats do not go far enough in explaining what might have been fake, especially when some in the news media unabashedly claim, "All that is fit to print."

The U.S. news media organizations have never been too careful not to let themselves become victims of clever hoaxes. On the eve of Halloween on October 30, 1938, when war drums were getting louder in Europe, the famed Hollywood actor Orson Welles created widespread panic in Grovers Mill, a small town in New Jersey, by broadcasting the hoax that Martians (based on H. G. Wells's *War of the Worlds*) had landed in the town. Some people rushed to their churches believing that the end of the world had come, while others got ready to fight the Martians. Radio was the major news source in those days, and people accepted as truth whatever they heard on the radio. Welles wanted to debunk that blind faith with his radio hoax, but instead he aroused the wrath of the people and Congress. Today the Federal Communication Commission would impose a criminal penalty on any station broadcasting a hoax today.[37]

While the Welles radio hoax fooled only the ordinary folks of New Jersey, Vanderford and his two collaborators with their cyberhoax succeeded in fooling major news-gathering organizations that feed most of the world news media. Although we do not know how people reacted to the fake cyberbeheadings and whether anything could have been done about it, especially if the website is beyond the jurisdiction of the United States, the mainstream media cannot not be too careful about what they report. To set the world on fire, you don't need six degrees of separation in the age of the Internet, where the fake and the real cohabit comfortably, as we saw in September 2012 in the Prophet Mohammed YouTube video that went viral, causing widespread protests in Egypt, Libya, and elsewhere. Under the cover of public protests in Libya, Al-Qaeda terrorists killed U.S. ambassador Chris Stevens and three other Americans in Benghazi.

NEWS: SIMULATION AND DISSIMULATION

Sometimes traditional news media are caught between what bloggers and independent online journalists offer them gratuitously and what is insidiously fed to them by official sources. In the process, journalism has been losing its character, trust, and nobility.

The courage to investigate the truth hidden under plausible surface appearances and to report it accurately, regardless of the consequences, is what makes journalism a noble profession, though at times a dangerous calling.

Wall Street Journal reporter Daniel Pearl was brutally murdered by Islamic militants in Karachi in 2002. No blogger or reporter from a small news organization with limited financial and political resources would have undertaken such a hazardous assignment, a fact that should not be lost when alternative methods of news gathering and dissemination are reviewed to save journalism from a slow death due to the emergence of new communications technologies and Internet platforms.

At the time of Pearl's murder, an intriguing question arose as to whether the American credibility in foreign countries, especially Muslim and Arab nations where the battle to win the minds of the people has been waged for a long time, would be enhanced through disinformation or truthful and accurate information. A case in point: A group called the Office of Strategic Influence, located somewhere in a dark corner of the Pentagon, was set up in 2001 to plan several initiatives to win friends and influence people, especially in the Muslim world, through what the *New York Times* reported as "the blackest of black programs to the whitest of white."[38] The office of black-and-white lies, simulation and dissimulation, was set up in the aftermath of September 11 to counter anti-American propaganda through information warfare, but it was closed in its proposed format as soon as its existence became public knowledge.

The strategy was to wrap up propaganda under the cover of genuine news so that foreign news organizations would swallow them up like hungry pups. What is good for the gander is also good for the goose; so American news organizations, too, might gulp down the propaganda in these days of ever-pressing deadlines and competition. But what would have prevented the AP Wire to pick up a story from Al Jazeera and repackage it as value-added news analysis? In the process of the supply chain of news gathering, news media unabashedly borrow, steal, and cannibalize each other all the time. Just Google a news story—all the sources say almost the same thing, as if they came from the same original source in spite of a clever turn of phrase here and there. One might say that the information warfare strategy of the U.S. defense department under the Bush administration was not dissimilar to that of *Playboy* that wraps up serious interviews in the bodies of nude women, conducting a kind of vagina monologue to beguile the reader. But *Playboy* attempts to appeal to the prurient interest of the reader, not to enlighten him. *Playboy* has little news credibility, though it titillates the overexposed sensibilities.

Such is the power of American cultural hegemony that most foreigners indiscriminately and ravenously eat up made-in-America products, including news and entertainment. It feels good to mimic the Americans, though it may not make any sense. Sometimes there is no choice. The Associated Press and Reuters, the two most dominating global news agencies with reporters and stringers in every nook and corner of the world, reveal the daily world to most of us. We trust they do their best to be impartial and objective, though this is a false hope because news gatherers and analysts see events through their own framework, their own network of assumptions, their own tinted lenses, though they do not intentionally plant stories to mislead their readers. In good faith, we suspend our disbelief and give the news stories provided by the news media the benefit of the doubt. The Pentagon's proposal, therefore, to simulate genuine news to propagate American views abroad through legitimate news and wire services as well as the Internet was a political land mine that would have eventually blown up in its face, and it also would have destroyed the credibility of the news distribution organization upon which newspapers, especially small ones that cannot afford their own reporters everywhere, depend for their daily feed of news and views.

Dark Journalism

There was another dark, intriguing proposal circulating in the media. The Office of Strategic Influence also planned to send pseudonymous email messages, via friendly strangers, to foreign public officials, journalists, and civic leaders advancing an American point of view and to attack anti-American governments. So if a person received email from some university professor in France or the United Kingdom, they might have wondered how much of it was genuine and how much from the Pentagon. When the story was leaked that the Office of Strategic Influence was planning to outsource the dissemination of false and misleading information to outside contractors, in spite of subsequent troubling denials by Bush administration officials, Defense Secretary Donald Rumsfeld, and Vice President Dick Cheney, some damage had already been done. In this atmosphere of simulation and dissimulation, honest journalists might be accused of collaboration with the Pentagon. Pearl's captors accused him of being an agent of the CIA as well as of the Mossad, the Israeli intelligence agency.[39] Pearl was an honest journalist, but how does one know that an American journalist is a genuine person, and not a spy?

Propaganda, disinformation, and psychological deception have been used as war weapons since man began killing man in organized wars, but using these tools against both friendly and unfriendly nations alike is not a strategy of persuasion. It is a strategy of disaster for journalists to cooperate with the Pentagon. The Pentagon did not need a shadowy organization like the Office of Strategic Influence, certainly not in the age of the Internet when Twitter, blogs, and cyberjournalists can penetrate a most closed society and expose its excesses.

In spite of all its faults, the U.S. culture of openness, the shining side of the aspirational society, is the greatest source of its influence. Even today American news media have the reputation of being the most courageous and the freest in the world. The best the White House and the Pentagon can do is to let freedom speak for itself in the mobile-technology-driven marketplace of ideas.

SOCIAL MEDIA AND ONLINE CIVIC JOURNALISM

Professional journalists occasionally, and rightly so, have been accused of ignoring authentic events such as accountability reporting, hyperlocal news, and seemingly small things that matter to ordinary people.

It happens because sometimes professional journalists are too busy with sycophantic reporting of the White House, obsessed with scoops and perpetually breaking news, investigating sex scandals of the high and mighty, pursuing or concocting exclusives to advance their careers, or just filling up the news hole for well-established news categories. Bias and prejudice do creep in the way a news story is selected and framed, but the worst happens when something is excluded from publication. The invisible hand of the marketplace imposes its own logic of inclusion and exclusion on the process of selection of what goes into news, analysis, and commentaries. It has been frequently said that freedom of the press belongs to those who own the media or their hirelings who broadly share their worldview (A. J. Liebling, "Freedom of the Press belongs to those who own one").[40]

Journalists, during their apprenticeship or training in journalism schools, assimilate the culture and ethos of established news organizations and unintentionally became supporters of the establishment. Public spaces where news happens are well known, and reporters go on reporting on events they are expected to report, thus routinizing news, boxing it in a "situation room," which could be done by computers in the future or could be even outsourced.[41]

News management has become the management of expectations. When Michael Jackson died suddenly on June 25, 2009, of cardiac arrest, the news media performed to fulfill the expectations of the people as if they were following the script that had already been written and tried in the case of Elvis Presley and Princess Diana, for example. The Michael Jackson news coverage became a celebratory ritual much like TV networks' periodic ritual *Inside the Obama White House*, beautiful but mindless extravaganza to build up networks' audience.

Seeking access to the White House and other hubs of political power or becoming a conduit of influence peddling for lobbyists has become a major interest for some journalists and news organizations, and it has an adverse impact on their credibility and trustworthiness. The *Washington Post* was caught red-handed in influence peddling when *Politico*, a journalism organization that collaborates and distributes its contents via several media platforms, exposed the *Washington Post*'s planned $2,500 dinner for exclusive guests, lobbyists, and their clients, who could meet with important beltway leaders in a friendly, nonconfrontational environment.[42] Access to influential journalists and news organizations in America has become commoditized. Hundreds of thousands of cyber and citizen journalists cannot do what traditional journalists can do to feather their own nests.

So what can a citizen journalist with a cell phone video camera and computer do to bring real news to the people?

One might say that untutored citizen journalists have their own news sense. They have something important to share with others, and the best of them expand and add value to traditional journalism. Many newspapers in their online versions incorporate citizen journalists' stories, thus becoming content aggregators. Audiences become coauthors and are invited to post their comments on news reports and analysis on the digital versions of newspapers and broadcasting stations. None of them probably think that they might one day become Edward R. Murrow or Walter Cronkite. But the thrill of serendipitous discovery, being at the right time and right place to record and publish, for example, an event like the safe landing of the US Airways plane on a flight from New York's LaGuardia airport when it was struck by a bird over the Hudson River on January 15, 2009, and instantly share it with the world is a great reward.

EMPOWERING CITIZEN JOURNALISTS

The power of the tech-savvy ordinary citizen to record events is being channeled and repurposed by the forces of the marketplace. On June 28, 2009,

YouTube started the Reporters' Center, an instructional site for training amateur citizen journalists how to become effective reporters. The blog Helping You Report the News challenges the budding citizen journalists: "Ever captured a natural disaster or a crime on your cell-phone camera? Filmed a political rally or protest, and then interviewed the participants afterward? Produced a story about a local issue in your community? If you've done any of these things or aspire to, then you're part of the enormous community of citizen reporters on YouTube—and today, we're launching a new resource to help you learn more about how to report the news."[43]

Embedded in this paragraph were links to clips of the earthquake in Sichuan, China, on May 12, 2008; a bicyclist assaulted by the New York Police Department on July 25, 2008; a Virginia Republican senator, George Allen, mocking and directing his election rally audience's attention to an Indian American whom he called *macaca*, a derogatory term (monkey), on August 11, 2006; monks protesting in Myanmar and an interview with a protester at a rally; and a local issue such as biodegradable bags, all shot and uploaded to YouTube by people with their cell phone cameras. The Center includes a how-to video instructing the citizen journalists to learn "how to prepare for an interview" from former CBS News reporter Katie Couric (now a successful afternoon talk show host competing with Ellen DeGeneres for the hearts, minds, and the purse of American women); how to be an investigative reporter from the legendary *Washington Post* journalist Bob Woodward; and why it's important for citizens to participate in the news-gathering process from Arianna Huffington. And not least, *New York Times* columnist Nicholas Kristof's video on how to report from a crisis area without getting shot.[44]

What could be YouTube's motive? To encourage people, one might guess, to move away from aimless chatter that fills most of space in social media and to do something meaningful that could be used journalistically, so that YouTube itself becomes a primary source of raw, unfiltered news, which could be consumed as such by the public or could be used by professional journalists for further investigation and structured into a proper news story—in short, to crowdsource news gathering and make it marketable. Citizen journalists could become unpaid electronic stringers, whose raw stories could be reframed and rewritten according to the needs of the established news organizations, a way of co-opting distributed journalism into a centralized, traditional news organization.

Exaggerated Fears

Edward Wasserman, Knight Foundation professor of journalism ethics at Washington and Lee University, wondered whether "the standards for online news be different from those of existing media."[45] For online news to become trustworthy and reliable, it must have the well-established standards of the traditional media, which include verifiable facts based on rigorous investigation and the separation of facts from opinion.

A traditional media reporter's job might begin with a conversation, but it ends with a story that tentatively answers *what*, *when*, *where*, *who*, *why*, and *how*; nonetheless, it remains open ended for future development and becomes part of the national discourse. Every news story, whether it originates from a highly reliable source such as the *New York Times* and the *Washington Post*, the *Wall Street Journal* or a tabloid such as the *National Enquirer*, creates a potential for discrimination, prejudice, or libel and therefore potentially it is no different from what citizen journalists might put online. So the temptation is to ask the question that when you see a news story about Iran, for example, in the *New York Times* and an Iranian news blog, whom would you trust? The *New York Times* is a branded news source, and in spite of the weapons of mass destruction (WMD) news fiasco and the fact that it has been losing its readers, its brand reputation has not diminished. We are less likely to distrust the *New York Times* in spite of Judith Miller and Jayson Blair than the Drudge Report and Daily Kos, though these blogs may be the first websites we surf in the morning.[46] We need another source to confirm what we hear from bloggers and cyberjournalists because their accounts might not have been vetted.

Wasserman considers unbranded news as conversational, tentative, and inconclusive, "in part because online news really mashes together two different kinds of information exchange: conversation and publication. Conversation is tentative, a mutual process of sharing and clarification where facts are elusive. Imagine a phone call about a neighborhood happening—'Here's what I know, what have you heard? The model of news-as-conversation is one that many online practitioners extol.'"[47] Professionally reported news is more than conversation because it has a shape and form even though, like conversation, it is open ended. At its best, news is an empirical observation that should challenge the reader: prove me wrong. But it is presented as authoritative even when it is in the making, its ephemeral value depending upon the credibility of the source, the news organization, or the informant.

Wassermann's fears are exaggerated when he says, "Digital media enable every e-mail, every tweet, every blog entry, to go global; every conversation can become a publication, every phone call a radio broadcast. And I think that's dangerous."[48] But who is listening and watching these billions of tweets, blogs, and personal broadcasts? And why would anyone trust them? That's why YouTube's venture to train citizen journalists in the rudiments of news gathering and storytelling is of great significance because someday they might become "the long tail" of traditional journalism and eventually get domesticated.[49]

As the traditional media gradually give up their print platform and move online or continue as a hybrid, the question of brand and trust will be the most important for them to face, especially when Google News and other news aggregators rank, at least at present, all news stories upon their clicks-based popularity. Regarding market-based journalism, Wasserman's admonishment cannot be ignored, especially when he writes that "a journalism that is deliberately used to nurture and harvest demographically desirable markets on behalf of advertisers—that, to me, is the latest frontier of corruption, and the one that we must be most vigilant about fighting."[50]

Nonetheless, the role of commercial speech, advertising, and marketing, as a source of innovation and stimulus for the marketplace of ideas, cannot be underestimated in a free society, as we shall discuss later.

NEWS MEDIA AND INNOVATION

One of the greatest innovations of the television age has been the development of the television "spot," the thirty-second commercial, which revolutionized not only marketing and how Americans fight elections through jingles, slogans, visuals, and dramatics but also how Americans think, shape, and bundle their ideas to influence and persuade others. This habit of mind that a lengthy discourse, or an argument, can be reduced to a pithy statement without losing its appeal or intellectual force found its expression in the Internet age in the 140-character tweet, whose power we have yet to realize.

From the development in mid-nineteenth century of the Penny Press that reduced formal, stylistic language to its natural simplicity to the twentieth-century television commercial with its mass appeal and now tweeting and texting, from simple grunts to precision thinking, the development of public and interpersonal communication has been remarkable.

On Twitter, the message, shorn of social formalities, becomes the real thing, or it could be nothing.[51] In Twitter the medium does not influence the message, contrary to what the media scholar Marshall McLuhan argued long ago that the meaning of a message is shaped and controlled by the medium that carries it.[52] The filtering effect of the medium in Twitter is marginal. Twitter is an exceptional medium, a unique mode of communication, in the sense that you can't hide yourself in voluminous long-windedness simply because Twitter does not permit it. If you can't put together your thoughts and emotions in 140 characters, the maximum length allowed in Twitter, try some other platform: a full-page ad, infomercial, a blog, or a soapbox to rant to the world.

The message in Twitter trumps the medium and thus has the potential to carry an awesome blow, like the impact of a direct force or a lethal injection. "Death to the dictator," "Marg bar dictator," street protesters shouted and tweeted ceaselessly in Farsi and English in Teheran in postelection demonstrations remonstrating against the vote rigging that reelected Mahmoud Ahmadinejad as Iranian president in the June 2009 election. In November 2008 in Mumbai, India, people with cell phones tweeted in real time about the ongoing Pakistani terrorist attack that killed 163 people at the Taj Mahal Palace Hotel and other places in the city, a tragic event that highlighted the extraordinary role the social media has begun to play in India's daily life. Ponder over the simplicity and directness of the tweet: "There's a plane in the Hudson. I'm on the ferry going to pick up the people. Crazy."[53] Janis Krums, a resident of Sarasota, Florida, who happened to be on a passenger ferry on the Hudson River, tweeted on January 15, 2009, with the photo of the US Airlines that had belly-landed on the Hudson, moments after it left New York's LaGuardia airport where the plane was hit by a bird or two. Before the news media crew of big media corporations could get to the scene, Americans were dazed watching the rescue of passengers, tweet by tweet, as hundreds of ferry passengers and nearby onlookers uploaded their pictures and videos of the plane bobbing on the Hudson.

Exceptional

Citizen journalists and activists all over the world are using Twitter, blogs, and social networking sites in myriad ways and for different purposes. From webcasting the July 2009 ethnic riots in China's Uighur province, Iran's authoritarian Mullahs crushing the street demonstrators, the overthrow of dictators in Tunisia, Egypt, and Libya, and the Chinese blind activist Chen Guangcheng's escape in April 2012 from his house arrest to seek refuge in the U.S. embassy in Beijing and the subsequent diplomatic drama, somewhere someone is tweeting what is happening in the world, whether it is political oppression, accidents, or personal grievances. We don't need Dan Rather, Bob Woodward, or Seymour Hersh to do that for us.

But tweeting has its hazards. The expanded freedom does not free us from the rule of law if we tweet publicly, because someone might feel hurt and file a defamation suit. It happened to Amanda Bonnen, a Chicago resident, who in a fit of anger tweeted against her landlord: "Who said sleeping in a moldy apartment was bad for you? Horizon realty thinks it's okay."[54] Amanda Bonnen was grouchy about a water leak that had occurred in March 2009 that made her apartment moldy, which the Horizon Group Management, the realty company, had not fixed. America is notorious for being a litigious society. Unsurprisingly, the Horizon Group Management in its defamation claim asked for $50,000 in damages, albeit the court found the tweet nonactionable as a matter of law.

Everyone is asking us: Follow me on Twitter. Journalists, senators, Hollywood celebrities, the pope, and President Barack Obama—everyone is urging us to follow them. Politicians and celebrities, who are too busy to tweet, hire shadow Twitterers, but the joy and thrill of doing it on one's own is so great that some get into hot water. This is what happened to rock musician Courtney Love. In early 2009 Courtney tweeted against clothes designer Dawn Simorangkir about overbilling her for custom clothing. In her strident, saucy, ungrammatical rag-tag outburst, Courtney bubbled, "oi vey don't f— with my wardrobe or you will end up in a circle of corched eaeth hunted til your dead."[55] In response to her rants, Simorangkir, the clothes designer, filed a suit against Courtney Love in Los Angeles Superior Court for defamation, invasion of privacy, and emotional distress.

The power of Twitter comes from its terse economy, its conciseness, its snappish brusqueness created by the weak ties of its users. And as it evolves to become a mass organizing tool as well as a global listening post for the collective brain of a local community, a nation under the boots of authoritarian rulers, or human rights violators, the consequences can be far reaching. It has become a textbook case as well as the stuff of legend as to how and why Barack Obama's 2008 presidential campaign was so successful against all odds. It was the adroit harnessing of the power of social media including YouTube, Twitter, Facebook, MySpace, BlackPlanet.com, LinkedIn et al., which enabled him to raise millions of dollars to fund his political campaign.

Most of all, his simple and direct message of hope and change through social media appealed to the younger, tech-savvy, mobile millennial generation. Since then his political campaign has kept the sites updated with the latest videos, photos, and economic and political messages. Even after occupying the White House, President Obama has continued his political campaign approach out of political necessity, especially in those days when he was struggling to have his massive health reform bill passed by Congress. Apart from bouncing from one town hall meeting to another, he asked the American people on his website Organizing for America to "Tweet Your Senator" and urge them to pass the health insurance reform measure. During the post-2012 election financial cliff crisis, the president went on twittering ceaselessly.[56]

Twittering, of course, is no substitute for political skills and statesmanship for dealing with the opposition that has its own political agenda. Nonetheless, when people tweet, can the powers that be, whether corporations, politicians, democrats, or autocrats, afford not to heed their voices? Think about what happened in South Korea when under pressure from the Bush administration South Korean president Lee Myung-bak lifted the ban on beef imports from the United States in April 2008, which initially led to protests on social networks, but eventually those protests swelled and spilled into the streets of Seoul.[57] Lee apologized and reimposed the partial ban. Today governments, authoritarian and democratic, are watching lips and listening to whispers in cyberspace. The challenge might come from anywhere.

The use of social technology differs widely from society to society, according to a report by a British research agency, TNS, that conducts an international survey covering sixty countries and seventy-two thousand people.[58] People in emerging digital markets use social media more than their counterparts in developed societies. "The research shows four out of five online users in China (88 percent) . . . have written their own blog or forum entry," according to the report. It remains to be seen how long the Chinese people, growing richer by the hour, aware of the freedoms Americans, Europeans, and other democracies enjoy, will continue to accept censorship when it has become possible for people to aggregate their weak ties into a mighty force. The force that overthrew authoritarian governments in Tunisia, Egypt, Libya, and Syria in a revolutionary movement, euphemistically termed the Arab Spring that began in December 18, 2010, is an example of how social networks generated weak ties that can become a massive revolutionary force.

THE POWER OF WEAK TIES

Anders Sandberg is a distinguished research fellow at the Future of Humanity Institute at Oxford University. He is neuroscientist and a futurist and a proponent of human enhancement through technology. Commenting on social networking, he said, "A fluid, global society might be much better off having a lot of weak ties and a few strong ones."[59] It is through weak ties that new information can be gathered serendipitously and innovative ideas can emerge. The noted sociologist Mark S. Granovetter in a classic research paper defined a tie and its strength as "a combination of the amount of time, the emotional intensity, the intimacy (mutual confiding), and the reciprocal services which characterize the tie."[60] He argued that people with large number of weak ties are more likely to spread new ideas and diffuse innovation than people with strong bonds who form a small, walled-in group or cohort where information is more likely to stay put than to leak out and diffuse.[61] We need small groups with strong ties to keep secrets, but it is through loose networks of weak ties that people share their common problems and aggregate ideas and, when the need arises, emotionalize them into mass civic actions. Journalists, too, initially use weak ties to pick up story ideas and leads, which they develop further by using their strong ties and confidential sources.

The Internet was created to break barriers and open doors. Its protocols and architecture were designed to enable information to leap across incompatible systems, replicate, and grow through weak as well as strong ties. The goal was to make information indestructible. The ancient library of Alexandria, Egypt, was burned in 48 BCE. It is gone forever. But the Library of Congress will be there forever once it is completely digitized and goes online. The original purpose of creating the Internet was to share and collaborate with people with whom trust was already established; for example, experts working in defense labs and top research universities, in short, to strengthen ties that already existed. But the Internet has evolved into a metasystem that enables social networking relations based on weak ties, just good enough for keeping up with social conversation not only with friends, neighbors, and grandchildren but also strangers. Most importantly, it is the ability to connect with strangers that is at the heart of the Internet revolution.

But people today are increasingly becoming dependent on social networks to enhance and manage relations in spite of serious concerns about the threat to the invasion of privacy and surveillance that weak ties facilitate. The corporate world does not know what to do with social media, perhaps due to the fear of industrial espionage.[62] In 2010 German automaker Porsche, for example, banned its employees from using social networking during office hours, fearing that information could be leaked through social networking sites, especially Facebook. Porsche is one of Germany's crown jewels. Foreign industrial intelligence agencies persistently use Facebook to contact company insiders to win their trust to obtain trade secrets and sensitive information. And then there is the overhanging challenge of WikiLeaks and the opportunities for openness it provides, which we will discuss later.

The emergence of social networks has uncovered in us a universal desire to share our information, albeit with tremendous consequences ranging from cyberbullying to innovative practices in business and politics. Sharing is nothing new. What is new is the freely available and easily accessible technology that makes sharing and collaboration possible from many to many, though not necessarily based on strong bonds. What are the consequences of so many people getting together not as close friends but as a floating mass of people with variable bonds of relationships ranging from casual to intimate? Clay Shirky, in one of his latest books, *Cognitive Surplus: Creativity and Generosity in a Connected Age*, says, "When you aggregate a lot of something, it behaves in new ways, and our communications tools are aggregating our individual ability to create and share at unprecedented levels of (everything) more."[63]

Millions of tweets can capture the attitude, wisdom, or panic of the crowd and predict the stock market, according to Indiana University researchers Professor John Bollen and his PhD student Huina Mao. They found a high correlation between the Twitter index and Dow Jones Index, giving them a prediction accuracy rate of 87.6 percent.[64] Social networking is releasing so much energy that we seem to be at a loss as to how to use it in the best possible ways; but those who know reap tremendous benefits. Tea Partiers, for example, have been turning the weak links of millions of social networking Americans into a massive, decentralized political force.[65] But the question is how much trust we can develop in social networks built on weak ties.

BUILDING TRUST IN SOCIAL NETWORKS

On the Internet it is possible to know whether you're a sheepdog or bulldog.[66] Or, whether a person is a dirty trickster; an older man trying to seduce a young girl; a gender-swapping man playing with big boys in a cave in a virtual world such as *Second Life*; a teenager posing as a medical expert. Once it was the *New York Times* that was a newspaper of record, touting the motto, "All the news that's fit to print." Now Facebook, Google, LinkedIn, and other search engines and social networking sites have become, by default and design, not only the nation's record keepers but also the keepers of surfers' personal history. The idea behind surf-and-search record keeping is to create patterns that anticipate the searcher's behavior and provide the information quickly the next time he looks for an item. Moreover, Google, based on searching patterns, could answer a hypothetical question, for example: Where should I invest my bonus, in mutual funds or a saving certificate? Google's CEO Eric Schmidt calls it "augmented humanity."[67]

Freedom in the digital age is becoming "curiouser and curiouser," as Alice would have wondered and lamented in Wonderland. The more freedom you have, the more you seem to be losing it, and you don't know what to do about it or how and where to seek protection, whether under the First Amendment or some digital trusted system. Unlike the traditional media, social networks are capable of creating a virtual environment that can be designed to be persuasive and can motivate people to disclose personal information and even change their social behaviors, another example of the power of weak ties. Social engineers and behavior manipulators, especially those who serve authoritarian governments, know the potential of the Internet for creating an illusion of boundless freedom only to entice people in invisible digital chains and perpetual surveillance. Can such a society under constant surveillance continue to be creative and innovative? Will Americans adapt to surveillance and continue to be tinkerers and innovators?

The next challenge for software programmers, whether they work for Google, Yahoo, or some authoritarian regime, is to build a network architecture that enables the designing of virtual environments to motivate people, for example, to buy, sell, invest; obey the supreme leader; or even induce good social and personal habits such as not to drink and drive and not to have unhealthy sexual behaviors. Computer codes do have the potential for persuasive and coercive power, almost the force of law, according to Harvard Law professor Lawrence Lessig.[68] In fact, programmers could bypass the government and, in a manner of speaking, take the law into their own hands. Or the government could also use programmers to stifle or create dissent. This is another warning about our posthuman, post-Internet future.

The rhetoric of software design, the persuasive code that entices, builds relationships, arouses and fulfills desires, and that keeps the users coming back has not been explored by news media organizations, including daily newspapers struggling to survive in the age of Craigslist and Google AdSense. There is a great fortune in developing codes that persuade the user to change his attitude, behavior, and actions, which is in the realm of advertising and could be used by newspapers as well. It is also in the realm of gene manipulation and designer drugs of the future.

The strength of social networking is its interactive immersion, its ability to respond and give instant feedback and gratification. Selective self-presentation makes it possible for people to open themselves up to others, which they would hesitate to do in face-to-face conversation for fear of contradiction and lack of control. The playfulness of the virtual environment of weak social ties, an environment of "be all you can be," creates a pleasurable experience, a sensuous flow, in which we feel in control of our environment that real life might deny us. J. B. Walther, a professor at Michigan State University, calls it "hyperpersonal," something socially more desirable than we are likely to experience face to face.[69] It stimulates the playfulness of fantasy partly to compensate for the absence of aural and visual information that gestures and voice create in interpersonal encounters.

Fantasy lowers our guards and makes cyberspace seductively persuasive—and dangerous. So many teenagers go astray in cyberspace because it lets them assume fake identities and gives them the freedom to pretend what they fancy themselves to be. Some of them become victims of conmen and predators, who, too, assume identities desirable to their teenage victims. An avatar can be very deceptive. And Google, Twitter, and Facebook keep digital impressions of what is going on in our hearts and minds. But at the same time, when we tweet a random thought, Google for a news item, Facebook with friends, or upload a YouTube video, we permanently leave an itsy bitsy part of ourselves in cyberspace. Over a period of time, these bits and bytes or digital droppings can be turned into profiles that can predict behaviors including buying, selling, investing, dating, mating, and perhaps who would be the next handgun- or assault-weapon shooter in a school, mall, or multiplex. The delete key, as has been said, is deceptive and illusory; someone might have already accessed the information and passed it on to others. Digital information goes viral so quickly that it is impossible to recall it.

A curse of the digital age, as American politicians have been discovering, is certainly a public blessing. Now no politician or corporate head can hide from us. The Internet has appropriated the central stage in our lives, replacing the local newspaper. Unlike switching off a news channel or unsubscribing from a newspaper, opting out of the networked world is no longer an option. Our digital past cannot be deleted or modified, but it could certainly enhance or haunt us. In the 2006 Senate race campaign, Virginia Republican senator George Felix Allen pointed at a person in the crowd, and mocked, "This fellow here, over here with the yellow shirt, macaca, or whatever his name is."[70] The racial slur did him in. And in 2010 when Tea Partier Christine O'Donnell tried to undo her 1990 remark that she had "dabbled in witchcraft" by issuing an ad, "I am not a witch," few seemed to believe in her naïvety.[71]

A Pew Internet & American Life Project showed that 51 percent of adults in the United States have uploaded more than one profile into social networks, which they keep monitoring and updating.[72] Most adult Americans use search engines to check information about themselves, perhaps because maintaining one's reputation and privacy has become difficult in the age of social networks. Today most Americans do online research about specialists, doctors, surgeons, and financial experts before they ask for their professional service. It is not uncommon for dating partners to do online research about each other before they get into serious relations.

The enthusiasm about social networking among the young and the old all over the world has been increasing steadily, and the main reason for its growing popularity, especially among the wireless mobile, is that it makes the users' lives easier. Besides, it creates associative or serendipitous possibilities where none existed earlier. Nonetheless, the American people are worried about the government and private companies collecting information about them when they are online. Data-sniffers do make us vulnerable on the Internet. If in a shopping mall someone watches or stalks you, you become alert and take some action, or maybe you choose to do nothing. In a digital mobile world, you don't know who is watching you and why, which creates a diffused sense of anxiety and consequently reduces trust in the system.[73]

The adult users' view of the Internet contrasts sharply, in many respects, with what American teenagers think about cyberspace. Teenagers love the mobile freedom and anonymity. For them it is a source of empowerment and a zone of unsupervised freedom. A Pew/Internet survey, "Teens and Social Media," reported that teenagers' "use of social media—from blogging to online social networking to creation of all kinds of digital material—is central" to their lives, with girls ahead of boys. "Digital images—stills and videos—have a big role in teen life. Posting them often starts a virtual conversation. Most teens receive some feedback on the content they post online," the survey reported.[74] Unlike adults who want the Internet to be regulated somehow, teenagers prefer the Internet to be left alone, lest its freedom be compromised. They are aware of the dangers of meeting strangers and predators online but feel confident of dealing with the situation on their own, a view that also finds expression in other reports about the Internet and teenagers.

Teenagers are also not as much concerned about surveillance as are adult users, which may seem a little puzzling. Teenagers' indifferent attitude regarding data mining and profiling is due to the fact that they have very little to lose in material terms; for example, credit card identity theft, financial blackmail, and bad credit. More importantly, it is this digital generation that the news organizations and media establishment have to take into account while developing new business models for delivering news and entertainment.

CAN SOCIAL NETWORKING CREATE SOCIAL CAPITAL DESPITE WEAK TIES?

In 1995 Robert Putnam evoked Alex de Tocqueville's 1830 visit to the United States during which the Frenchman had observed the American propensity for clubbing together in myriad neighborhood associations, which created civic norms and social trust and built social capital. Putnam lamented that America's social capital was declining due to people's diminished engagement in public and civic affairs. The American people did not care much for voting during elections; union membership was declining; participation in civic and fraternal organization membership had fallen, and so on. More Americans are bowling today, he said, but they are bowling alone.[75] Bowling alone does not create social bonds and build social capital. AARP has thirty-three million members, but that cannot replace a close-knit local organization, the basis of civic society and the foundation of democracy. In order to restore civic engagement and civic trust, Putnam said, restoring social connectedness should be the nation's top agenda. But whether such a small-town world of civic groups based on strong ties and cultural self-similarities can be reestablished in an increasing multicultural and highly mobile digital America is rather doubtful. Perhaps the sports metaphor "bowling" is not the best way to understand the complexity of American society today.

Nonetheless, the Internet and social networks have created opportunities for new forms of social connectedness, though with weak ties. The larger the group, the weaker the ties, especially when members are geographically dispersed and mobile. And that brings us to Malcolm Gladwell's article in *The New Yorker*, "Small Change: Why the Revolution Will Not Be Tweeted."[76] He argues that social media cannot bring about a massive, fundamental transformation in society, as happened during the civil rights movement of the 1960s, which "happened without email, texting, Facebook, or Twitter." That is true, of course, but what would have been the movement's fate without television and newspapers? Gandhi's nonviolent movement for India's independence did not have an altogether unsympathetic press in Britain, the United States, and Europe.

Social networking cannot instigate social activism, but it can hasten it by awakening dormant bonds and strengthening weak ties. Gladwell says, "Activism that challenges the status quo—that attacks deep rooted problems—is not for the faint of heart," because it requires structured hierarchy, command

and control, strong ties, and commitment.[77] These are not the attributes of social media, which is built on loose and casual relations. Unfortunately, Gladwell is keeping in mind old models of political and social revolutions led by Lenin, Gandhi, Mao Zedong, and Martin Luther King Jr., for example. The age of social media requires a new generation of social thinkers and activists who know how to transform weak ties into strong bonds whenever needed; for example, like the people who used social network–enabled weak social ties in 2010 to 2011 to organize the Tea Party and the Occupy grassroots movements that significantly impacted politics in the United States.[78]

The Protestant Reformation, which began with the nailing of Luther's ninety-five theses in 1517 and ended with the Treaty of Westphalia 1648, did not have a single command-and-control hierarchical structure that Gladwell talks about as indispensable to a movement. Nor did a single leader or a hierarchical organization start or direct the Age of Enlightenment. The Protestant Reformation, which was neither unified nor centralized, broke the hegemony of the Roman Catholic Church, liberated the European mind, and eventually ushered in the Age of Reason and Enlightenment. Europeans reclaimed the ancient Greek spirit of rational inquiry, and one might say it gave birth to the Bill of Rights, especially the First Amendment—the greatest human utterance ever uttered, perhaps.

It is difficult to say what new media and social networking sites will do to us, but let me share a recent thought experiment I conducted with my students. I asked them to imagine what would have happened if the German people had access to social networking, Twitter, Facebook, WikiLeaks, and more during the Third Reich, Nazi Germany. All my students gave only one answer: The Holocaust would not have happened. Imagine the possibilities of empowerment, when mobile technology enables people to open up the darkest secrets of the government! Let's celebrate the whistleblowers among us.

WIKILEAKS'S CHALLENGE

Benjamin Franklin once said, "Three may keep a secret, if two of them are dead."[79] Of course, one of them could be a kitchen maid, a valet, or perhaps a butler, as happened in the case of the Vatican where Paolo Gabriele, the butler of the former pontiff, Pope Benedict XVI, was arrested in May 2012 for leaking secret documents regarding the Vatican Bank, whose president, Ettore Gotti Tedeschi, was unceremoniously booted out.[80] These documents leaked to journalists were published in a book that gave details of the sordid power struggle, political intrigues, corruption, and crony capitalism at the highest level in the Catholic Church, raising questions about Pope Benedict XVI's ability to govern the Holy See, keeping also in mind the sex scandals that have dogged the Roman Catholic Church for a long time.[81] Pope Benedict XVI resigned in February due to health reasons, but the true story will come out gradually. In the digital age the Vatican has no walls that can't be breached.

Whether it is the Catholic Church, Guantanamo Bay prison, Wall Street bankers, or how China's Foxconn manufactures Apple's iPads, we need to know and have the right to know the facts. More than that, now we know how to seek information, thanks to WikiLeaks and other collaborative open source organizations. WikiLeaks should leave no doubts in our minds that the world that once safeguarded our confidences and secrets is disappearing. But secrets have never been totally safe. In fact, in the United States, whistleblowers are regarded as heroic and are protected and rewarded by state and federal laws.

Let's consider a few cases as to how whistleblowers have improved the quality of life in the United States. In 1966 Peter Buxtun of the U.S. Public Health Service exposed the Tuskegee Syphilis Experiment conducted between 1932 and 1972 on six hundred poor African Americans in Alabama. In a ceremony held in the White House on May 16, 1997, for the surviving participants of the Tuskegee Syphilis Experiment, President Bill Clinton offered a belated formal apology: "What was done cannot be undone. But we can end the silence. We can stop turning our heads away. We can look at you in the eye and finally say on behalf of the American people, what the United States government did was shameful, and I am sorry . . . To our African American citizens, I am sorry that your federal government orchestrated a study so clearly racist."[82]

In 1971 Daniel Ellsberg of the U.S. State Department leaked the Pentagon Papers, a secret historical account of the origins of the Vietnam War, to the *New York Times* and the *Washington Post*, revealing the deceptive practices of the Kennedy and the Johnson administrations and the generals. In 1972 W. Mark Felt of the FBI, who until 2005 was known as Deep Throat, leaked information about President Richard Nixon's involvement in Watergate to *Washington Post* reporters, Bob Woodward and Carl Bernstein. The Watergate scandal eventually led to the ignominious resignation of the president, and later on under the Ford administration to the enactment of the Government in the Sunshine Act of 1976, strengthening the Freedom of Information Act of 1966.

In 1996 Jeffrey Wigand, who was vice president for research and development at a tobacco company, Brown & Williamson, told the CBS news program *60 Minutes* that the tobacco company purposely manipulated the level of nicotine in cigarettes to cause addiction in smokers.[83] In 2002 Cynthia Cooper of WorldCom exposed corporate financial scandal and was named *Time*'s People of the Year in 2002 along with Sherron Watkins, who had exposed corporate financial scandal at Enron.[84]

Since its launch in 2006, WikiLeaks, founded by Julian Assange, has revealed more than a million documents. Beginning in February 2010 WikiLeaks began releasing top-secret, classified cables that were received by the U.S. State Department from its consulates, embassies, and diplomatic missions around the world. The classified cables, covering a period between December 1966 and February 2010, comprised diplomatic analysis from various world leaders as well as the U.S. diplomats' assessment of host countries and their top officials. The so-called Cablegate is credited with being an important contributing factor and a catalyst for the beginning of the Arab Spring that started with the Tunisian revolution, followed by Egypt, Libya, and the ongoing struggle in Syria and other Middle Eastern countries.[85]

This is the age of WikiLeaks. It is very unlikely that we can protect our precious secrets from someone who is driven by the ideology of openness or obsessed by the moral claim that all confidential and secret negotiations, deliberations, and discussions must be exposed in the public interest. That has been the case with Bradley Manning, U.S. Army intelligence officer in Iraq, who is under trial for leaking not only classified documents but also the horrific Iraq Collateral Murder video.[86]

Timothy Garton Ash is a British historian and professor at Oxford University, and author of several books and a columnist for the *Guardian*. He wrote in the *Guardian*, one of the newspapers that received a cache of more than 250,000 diplomatic cables from WikiLeaks.org, that "it is the historian's dream. It is the diplomat's nightmare. Here, for all to see, are the confidences of friends, allies and rivals, garnished with American diplomats' frank, sometimes excoriating assessments of them . . . a multi-course banquet from the history of the present."[87] Writing about the history of the present is the job of journalists who work for established news organizations and who now feel threatened by the rise of decentralized and distributed new media such as WikiLeaks and the rest of the blogosphere. Historians usually wait for thirty years before they get access to such important documents as have been released by WikiLeaks. According to an original estimate, the released cache of diplomatic cables contained more than 250,000 documents, which by some estimates contained about half a billion words.[88]

Making Sense

Even if we have access to such a huge treasure trove of information, it is impossible for most of us to digest the information and make sense out of it. That's one reason that WikiLeaks approached some of the most important newspapers in the world to sift through the documents and release them according to their own news judgment. What the public remembers, based upon newspaper reports, are nothing but some of the most embarrassing tidbits. The diplomatic cables go back as far as 1966, but most of them are of recent origin, covering the period since the Iraq War.

Most of the leaked diplomatic communication was about security and counterterrorism issues that have captured American foreign policy experts' attention for more than a decade. In one of the telegrams, attributed to former Secretary of State Hillary Clinton, it was suggested that American diplomats should behave like low-level spies; that they should go around to snoop credit card and frequent flier information of UN diplomats to track their movements. Other concerns included how North Korea aided Iran on its weapons program; South Korean and U.S. diplomats' discussion about what to do if North Korea collapsed; a cable about Pakistan's nuclear stockpile in danger of being seized by Islamist groups; Israel's worries about the Iranian nuclear program; and the revelation that Saudis, too, have tremendous fear about Iran's nuclear programs.

Another document revealed that King Abdullah of Saudi Arabia asked the United States to "cut off the head of the snake," meaning that Iran's nuclear program should be wiped out. Yet other cables quoted a senior Saudi official warning that if Iran was not stopped then other Arab states might develop their own nuclear weapons. The former Russian prime minister and now president, Vladimir Putin, was called an "Alpha Dog" in a cable. The Afghanistan president Hamid Karzai was driven by paranoia. German chancellor Angela Merkel was called "Teflon Merkel," a person who would avoid risk and is seldom creative. A voluptuous, blonde Ukrainian nurse always accompanied former Libyan leader Muammar Qaddafi. In another cable a diplomat wrote that the Italian prime minister at that time, Silvio Berlusconi, appeared increasingly to be Russia's mouthpiece after he received lavish gifts and energy contracts from Russia.

Some of the cables contained information that had been known to the American public for some time; for example, U.S. diplomats offered many countries incentives, such as meeting with President Obama in the White House, and also money, millions of dollars in some cases, if they accepted detainees from Guantanamo Bay prison. A cable mentioned that China's Politburo directed that Google's computer system be compromised.

There was a general fear at that time that the disclosure of cables, which were supposed to be classified for twenty years, might have a chilling effect upon ongoing diplomatic conversation and negotiation that American diplomats were conducting with their counterparts; the revelations might create tension and spoil the relationship between foreign diplomats and the United States.

Of course, the public is very much interested in knowing how U.S. diplomacy works, but at the same time it must be acknowledged that some level of confidential information is important to conduct foreign policy. But the most important question is: Can diplomacy be conducted if everything becomes transparent? Or, how should diplomacy be conducted in the age of WikiLeaks? At a Pentagon briefing on November 29, 2010, Robert Gates, former Defense Secretary, gave a balanced view:

> Now, I've heard the impact of these releases on our foreign policy described as a meltdown, as a game-changer, and so on. I think those descriptions are fairly significantly overwrought. The fact is, governments deal with the United States because it's in their interest, not because they like us, not because they trust us, and not because they believe we can keep secrets. Many governments—some governments—deal with us because they fear us, some because they respect us, most because they need us. We are still essentially, as has been said before, the indispensable nation. So other nations will continue to deal with us. They will continue to work with us. We will continue to share sensitive information with one another. Is this embarrassing? Yes. Is it awkward? Yes. Consequences for U.S. foreign policy? I think fairly modest.[89]

The transparency enabled by WikiLeaks and other similar organizations is certainly challenging the government with difficult questions as to how the United States conducts its foreign policy and pursues its global military objectives. There is no gainsaying the fact that the disclosure of secrets in some cases might damage the functioning of responsible governance. But the government and corporations today will have to work on the assumption that eventually secrets will be out, and when they do come out, they should consider what do about the consequences, whether they concern military and foreign policy, diplomacy, or domestic problems.

The most salient feature of the age of WikiLeaks is that today it is impossible for anyone to have total control or maintain their monopoly over critical information. The monopoly of the traditional news media for gathering information and presenting it to the public in their own framework has been broken. Bradley Manning, we should keep in mind, is a man with strong moral convictions who believes that what the government was doing in Iraq and Afghanistan was wrong and deceptive. And he just could not do without disclosing classified information regardless of the consequences, just as other whistleblowers have been doing. Manning has offered a guilty plea to some charges. It is not that WikiLeaks is indestructible because the whistleblowing code and the architecture that ensures bulletproof anonymity to whistleblowers are available and anybody can build another WikiLeaks. It is nevertheless time to think again about official and corporate secrets, their classification system, and the need to do so.

With WikiLeaks, Facebook, YouTube, Twitter, and cell phones, Cloud computing, and Big Data the burgeoning age of information makes it extremely difficult for any secret to be maintained for a really long time. So the biggest challenge for diplomacy or the government is: Can we do with fewer secrets? It's going to be difficult for the government to make a policy announcement but not to act upon it.

The Internet and organizations such as WikiLeaks will empower people to confront problems in a shared manner and also expose public corruption, as we saw in the case of the recent Vatican crisis. WikiLeaks has provided us with a blueprint for keeping the government as well as global institutions and corporations open and transparent, a most prominent characteristic of the networked age that will get better as traditional news media organizations, too, incorporate the concept and the architecture into their websites.

New Challenges

The challenge to the authoritative voice of the traditional news media and its claim to serving the public interest through investigative reporting, opinion-making, and agenda-setting functions, comes from various sources, including technologically empowered lone citizen journalists and bloggers, news aggregators, and opinion websites, but not least from the emerging new online public interest information organizations such as Cryptome and WikiLeaks that provide bullet-proof anonymity to whistleblowers and leakers of classified documents.[90] It is the challenge of the ever-increasing and growing environment of news ecology, unencumbered by advertising, that we have to keep in mind while evaluating how indispensable is the old traditional media for the free flow of information, openness and transparency, democratic self-renewal, and the culture of innovation.[91]

The traditional news media organizations are indispensable in the sense that they provide a frame of reference and selective perspective and assessment to enable us to understand what is happening in the world, a world which for most of us is too much. We cannot have thousands and thousands of citizen journalists uploading video clips and views online without someone aggregating them and putting them into the proper form and perspective.

We should keep in mind that it is the form and the shape of the news that generates sense and sensibility in the audience. It is the well-constructed news story appropriately placed with pertinent headlines and proper focus that gives us a sense of what is happening in the world.

We should visualize the emerging news ecosphere as a spiral galaxy. At the core of this evolving news ecosphere are the traditional news media, highly established, branded newspapers such as the *New York Times*, the *Washington Post*, the *Guardian*, the *Times* (London), and the *Wall Street Journal* as well as other news establishments with powerful political and corporate connections. Of course, the dominance of wire services such as Associated Press and Reuters and many other global news agencies in the news-gathering process cannot be underestimated: they are the eyes and ears of the world. From their raw sources, from thousands and thousands of reporters, stringers, and collaborators scattered all over the world, the news-gathering organizations shape the information that is further processed and value added by the traditional news media in order to make the events meaningful to us.

In this environment of multiplication of news sources it is all the more important that we should have someone skilled, accomplished, and competent who can guide us through the endless flood of information; people who can digest, recapitulate, and synopsize for us what is happening. The question is whether the traditional news media will be able to meet the challenges of alternative digital platforms, ceaseless news aggregators, bloggers, and global whistleblowers such as WikiLeaks, Cryptome, and other collaborative open source organizations, organizations without national headquarters, without brick-and-mortar buildings, organizations that exist only in cyberspace.

In the past the traditional news media have responded to challenges and evolved. For example, newspapers embraced the telegraph in the nineteenth century, which apart from speeding up news also transformed a news story into an inverted pyramid form with a synoptic headline and a lead.[92] The telegraph also necessitated and facilitated the formation of the Associated Press, the first news cooperative of its kind. In the course of time newspapers were transformed with the innovation of the Penny Press—inexpensive, advertisement-supported, tabloid-style newspapers for the masses—and the development of steam-powered printing, advancements in photography and offset lithography, and now digital printing.[93]

At the beginning of twentieth century the rise of electronic media, first radio and then television, once again challenged the print news media. Gradually various media, print, radio and television, and later cable settled into a dynamic system that increasingly enhanced the news environment. Radio, television, and cable did not replace the old print media; rather, they cooperated, fed upon each other, and formed a new media environment. But today all news media organizations in this changing news environment have the same setup in the sense that at the heart of this dynamic system is an organization of professionally competent people who look at the world through certain well-defined categories of news, desk, and beats. Whether it's the White House, diplomacy, finance, courts, police precincts, local news or sports, all pigeonholes are filled up to fulfill the expectations of people.

The system enables top-to-bottom, mass-mediated communication: a group of well-organized, highly trained people look at the world through the Associated Press, Reuters, and other news agencies and collect information based upon the perception of what news should be; thus the news is framed to be given to the American people regularly on a platter. It is this core organizing principle of news gathering and distribution that a handful of people could be trusted to look at the world and frame the information and give it to the audience that is being challenged by the rise of the socially networked news media. The Google News algorithm, for example, does not attach any value to a story, whether it comes from the *New York Times*, the *Washington Post*, CNN, or Fox. Its present story-ranking order is very different from that of the traditional news organizations, which they claim is based on sound editorial judgment.

Of course, we have to keep in mind that the reputation of the traditional media depends upon investigative reporting for which they need to cultivate sources. And some of those sources are confidential, as was the case with the Watergate affair. But WikiLeaks has made nonsense of this kind of relationship. WikiLeaks presents the greatest challenge to traditional media's investigative role in society.

How are the traditional media going to deal with an organization such as WikiLeaks? If one disappears, others will rise. The traditional media can survive only by collaborating with social media organizations. Journalism is not dying. The newspaper will not die.[94] They will become collaborative and interactive multimedia platforms, where the audience will be participants rather than passive consumers. But the challenge of what is news and what people need to know will continue to be crucial, because the distinct boundaries and contours of journalism, as Michael Schudson says, "which just a few years ago seemed relatively clear, and permanent, have become less distinct, and this blurring, while potentially the foundation of progress even as it is the source of risk, has given rise to a new set of journalistic principles and practices."[95]

Chapter Four

Commercial Speech and Innovation

THE SOFT POWER OF ASPIRATIONAL BRANDS

Is it a world to hide virtue in?
—Sir Toby Belch in *Twelfth Night*

Free speech and free flow of information permeate the marketplace with dynamic energy that propels business enterprises to compete for customers through commercial messages in various media.

Advertisement fulfills needs, stimulates wants, creates desires, expands the marketplace, and encourages businesses to innovate for competitive advantage. Under the watchful eyes of regulators and the scrutiny of the free press, advertisement enables people to make their economic choices that not only meet their budgetary needs and taste preferences but also increase economic activities.[1]

In a landmark case Justice Harry A. Blackmun said, "So long as we preserve a predominately free enterprise economy, the allocation of our resources in large measure will be made through numerous private economic decisions. It is a matter of public interest that those decisions, in the aggregate, be intelligent and well informed. To this end, the free flow of commercial information is indispensable."[2] To enjoy the plenitude and variety of goods and services offered by competing businesses and their persuasive commercial messages, people work more, earn more, and consequently live a dynamic life. As a handmaid of the free marketplace, advertising studies human motivations and behavior and how to sway them.

Although the First Amendment gives commercial speech substantial protection today, it is regulated to safeguard against unfair competition and to protect the consumer from deceptive advertising and unsustainable claims. Advertisement is a supreme example of speech that spurs action and the concomitant consequences; therefore, it is a category of speech that is to be watched carefully. Corporations are required to disclose the composition and origin of their products, a forced transparency that makes products safer in the United States. Moreover, news media continuously scrutinize corporate claims regarding their products and services, thus creating a culture of corporate accountability and social responsibility.

Public scrutiny, whether it is done by competitors or consumer activists, stimulates incremental innovation in a challenge-response environment. The open marketplace is not a steady state of equilibrium where a business can claim a permanent niche. There is always a challenger waiting in the corner who has a better product and an alluring message to persuade the consumer. To stay competitive or to beat the competition, businesses are always in search of disruptive innovations or killer apps.

Status Change

For a long time the American people believed that the function of the First Amendment guarantees of freedom of speech and the press was primarily to keep a check on the government. Commercial speech was thought to be self-serving and unworthy of First Amendment protection. It took more than a century to liberate commercial speech from the shackles of doubt it had created in the minds of the American people, who dreaded its excesses, false assertions, and deceptive claims.

Beginning in the early decades of the twentieth century when the U.S. Supreme Court began to consider the boundaries of political speech, the nature and the status of commercial speech could not have been ignored for long. As commercial speech found its rightful place in the economic and cultural life of the nation under the First Amendment, marketing innovation turned the United States into a dynamic consumer society. Mass consumption generated by advertisements stimulated a variety of other economic activities, adding to the diversity of jobs.

Few other forms of communication spur as much innovation as does advertisement because advertisers seek to understand people in order to persuade them and change their behavior. By arousing aspirations, by transforming wants and desires into compelling needs, it is commercial speech that is mostly responsible for transformative or disruptive innovation.[3] But every disruptive innovation requires acceptance and adoption by consumers, whether as individuals or businesses, which requires persuasion. The ultimate goal of commercial speech is persuasion through attraction.

Commercial speech—advertisement, promotion, and marketing communication—is a notable hallmark of American expressive behavior, which is not only the foundation of American consumer culture but also it is an activity that instigates social changes and stimulates innovation and creativity in myriad ways. Commercial speech as a handmaid of the marketplace needs large public spaces and multiple platforms to seek and persuade people, and in the process, no doubt, it has a propensity for excess. To control the cultivated exuberance and extravagances of commercial speech and its terrific transformative power, we need not only reasonable regulatory regimen but also an uninhibited press, whose freedom from governmental control is ironically underwritten mostly by advertisement revenues. The cohabitation between commercial speech and free press creates dialectical tensions and ethical debates in American society, best dramatized by the HBO drama series, *The Newsroom*, in which a group of high-minded journalists "set out on a patriotic and quixotic mission to do the news well in the face of corporate and commercial obstacles."[4]

Since the purpose of commercial speech is persuasion by using all available means, it tends to create fear, which in the days of snake-oil peddlers and instant-cures quacks was best captured by the aphoristic warning *caveat emptor* (let the buyer beware). But as American society started to evolve into a consumer society, especially from the beginning of the twentieth century, the diffused buyer-beware fear and consumer helplessness in the marketplace were mitigated by Congress that put in place a regulatory infrastructure for consumer protection, the bulwark established by the Federal Trade Commission, Food and Drug Administration, Federal Communication Commission, Securities and Exchange Commission, and the U.S. Postal Service.

As state and federal regulatory agencies began to scrutinize and regulate commercial speech in the interest of consumer safety, periodic challenges by enterprising businessmen, professionals, editors, the ACLU, and others as to whether the First Amendment was applicable to commercial speech began to hit the gates of the U.S. Supreme Court. The Supreme Court, with its ruling in several commercial speech cases from 1940 onward, started to ease, albeit haltingly, restrictions on commercial expression; nonetheless, in the course of time, it progressively brought up commercial-speech status to almost the level of political speech. And as trade and commerce increasingly rose to take up the center stage in American society, the Supreme Court could not ignore the political importance of commercial speech, the political economy of advertisement, and the increasing gospel of wealth that was being preached in the twentieth century.

Challenging Authority

The first significant challenge as to the rightful place of commercial speech in American society came from an adventurous and resourceful showman, F. J. Chrestensen. In 1940 Chrestensen bought an old navy submarine in Florida and hauled it to dock at New York port on the East River across from the Statue of Liberty in order to turn it into an exhibition piece so that people could see how a submarine worked from the inside. The goal was not to educate people but to make money. To attract visitors to the submarine, Chrestensen distributed leaflets in the city to promote his enterprise, but unfortunately he violated the city's antilittering code that disallowed the distribution of handbills of a commercial nature, albeit one could distribute political and informational handouts.

Not deterred by legal niceties, Chrestensen printed another set of leaflets that contained a political protest against the high-handedness and the unfairness of the city's ordinance on one side and the commercial message inviting people to visit the submarine on the other. The city authorities were not persuaded by his innovative gimmick and banned the distribution of handbills. The stakes were too high for Chrestensen, but he did not quit his quest quietly. He obtained an injunction from the federal district court against the enforcement of the ordinance, which was affirmed by the federal appeals court. The city police commissioner, Lewis J. Valentine, took the case to the U.S. Supreme Court. The Court ruled that the First Amendment did not give protection to "purely commercial advertising" from government regulation.[5]

Historically the importance of the Chrestensen case does not lie in the fact that the Supreme Court affirmed the subservient status of commercial speech; rather, the case is significant because beginning with the First World War and the Espionage Act of 1919, Americans had begun to test the boundaries of the First Amendment in all their activities. The state could regulate advertisement on the grounds that it is a vehicle for selling a product. But what if advertisement were used as a protest platform, or if it sold a political idea! In a 1964 landmark case *New York Times v. Sullivan*, the Court ruled that political ideas expressed as advertisement have the full protection of the First Amendment, even if they contain inaccuracies and falsehood, so long they are not intentional or show reckless disregard of the truth. Although it is difficult to say that the *Sullivan* case affected the Supreme Court's view about commercial speech, in the mid-1970s the Court started to reconsider its position on the importance of commercial speech in the daily lives of the American people.

In 1971 the *Virginia Weekly* published an advertisement informing its readers in Virginia, where abortion was illegal, about the availability of abortion services at affordable rates at the Women's Pavilion and several accredited clinics and hospitals in New York City, where abortion was legal. Since under Virginia state law it was illegal to advertise for abortion services, Jeffrey C. Bigelow, the *Weekly*'s editor, was convicted for misdemeanor. The Virginia Supreme Court rejected Bigelow's First Amendment protection claim and ruled that the message was of "purely commercial nature." But on appeal the U.S. Supreme Court reversed his conviction ruling that the controversial advertisement "contained factual material of clear 'public interest.'

Portion of its message . . . involve the exercise of the freedom of communicating information and disseminating opinion."

It was the first halting step in recognizing that if an advertisement contained truthful and factual information that mattered to the public, it deserved First Amendment protection. Delivering the Court's opinion, Justice Blackmun's statement included the newsworthiness of the commercial message that also had a constitutional dimension, especially women's rights:

> Viewed in its entirety, the advertisement conveyed information of potential interest and value to a diverse audience—not only to readers possibly in need of the services offered, but also to those with a general curiosity about, or genuine interest in, the subject matter or the law of another State and its development, and to readers seeking reform in Virginia. The mere existence of the Women's Pavilion in New York City, with the possibility of its being typical of other organizations there, and the availability of the services offered, were not unnewsworthy. Also, the activity advertised pertained to constitutional interests. . . . Thus, in this case, appellant's First Amendment interests coincided with the constitutional interests of the general public.[6]

This was the first time that the Supreme Court correlated commercial speech with women's rights and social development. By banning abortion ads, Virginia was shutting the doors on women to explore what Steven Johnson (and evolutionary biologist Stuart Kauffman) call *associative and adjacent possibilities*.[7] If commercial speech opens new avenues, let it be free. The judicial thinking that commercial speech needed greater protection than it had been hitherto given was further confirmed in a pivotal 1976 case, *Virginia State Board of Pharmacy v. Virginia Citizens Consumer Council.* The issue dealt with the First Amendment right of pharmacies to post the prices of prescription drugs that the Virginia law had banned, the law based on the assumption that aggressive advertising by professionals such as pharmacists would lead to unhealthy competition and a decline in standards.[8]

A Virginia consumer and two nonprofit consumer groups challenged the law, claiming that the First Amendment protected consumers' right to receive information provided by pharmacists. The question before the Supreme Court was whether the First Amendment protects the right to advertise a legal product and whether consumers have the right to receive the information even if it is purely commercial. Freedom of speech presumes a speaker willing to listen to the message. As Justice Blackmun, writing for the majority, said, "But where a speaker exists, as is the case here, the protection afforded is to the communication, to its source and to its recipients both."[9]

But simply because information is purely commercial in nature and the advertiser's motive is economic profit does not necessarily deprive the message of its First Amendment protection. Society has paramount interest in the free flow of information, even if it concerns pharmaceutical products. Justice

Blackmun elucidated that the interest of a person with limited income in the information about comparative product pricing "may be as keen, if not keener by far, than his interest in the day's most urgent political debate. . . . Generalizing, society also may have a strong interest in the free flow of commercial information. Even an individual advertisement, though entirely "commercial," may be of general public interest. . . . Advertising, however tasteless and excessive it sometimes may seem, is nonetheless dissemination of information as to who is producing and selling what product, for what reason, and at what price."[10]

In order to protect consumers the state may not suppress information offered by competing pharmacies, particularly the kind of information that is not unlawful. But if pharmacists could advertise their lawful products, why couldn't other professionals, for example, lawyers, do so? In *Bates v. State Bar of Arizona*, the Supreme Court considered the outright suppression of free speech by lawyers advertising their services. The state justified its punitive ruling on the ground that advertising would diminish the legal profession's traditional service emphasis, raise the cost of litigation, and turn society into a litigious one. Applying the Virginia *Pharmacy Board* reasoning, the Court ruled that while misleading and deceptive advertisement could be penalized, accurate informational announcements regarding the availability of legal services may not be banned.

COMMERCIAL SPEECH AND THE PUBLIC INTEREST

But what if truthful and accurate information hurts a larger public interest that the state is obligated to protect? In a decentralized judicial system such as the United States, what standards should lower courts follow? The Supreme Court confronted this question in a landmark case *Central Hudson Gas & Electric Corp. v. Public Service Commission of New York*, under which the Court established a four-part test in order to determine the government's compelling interest in suppressing commercial speech.

The case arose out of the contingencies of the Iranian Revolution of 1979 and the consequent oil crisis that led to rising oil prices and long lines at gas stations, reminiscent of the 1973 oil crisis when Arab countries of OPEC had imposed an oil embargo in response to the United States' support for Israel during the Yom Kippur War. To support the national energy conservation policy of the Carter administration, the Public Service Commission of New York had imposed a total ban on promotional advertising by electric utilities. Was the blanket ban constitutional even if the commercial speech content was not deceptive and there was no misrepresentation? That was the question before the U.S. Supreme Court.

Delivering the opinion of the Court, Justice Lewis F. Powell Jr. affirmed what the Court had already recognized in previous commercial speech decisions as "the 'commonsense' distinction between speech proposing a commercial transaction, which occurs in an area traditionally subject to government regulation, and other varieties of speech."[11] In order to help the lower courts to decide whether a commercial speech act has exceeded its limit or not, the Court established a four-part test:[12]

1. At the outset, we must determine whether the expression is protected by the First Amendment. For commercial speech to come within that provision, it at least must concern lawful activity and not be misleading.
2. Next, we ask whether the asserted governmental interest is substantial.
3. If both inquiries yield positive answers, we must determine whether the regulation directly advances the governmental interest asserted, and
4. Whether it is not more extensive than is necessary to serve that interest.

Applying the test, the Court recognized that the advertisement pertained to a legal activity and was not deceptive. Considering the international oil situation and the U.S. dependence on foreign oil, the government had a substantial interest in conserving energy, and the ban would advance the government interest since there was enough evidence that promotion increased sales. But the Court faulted the government on the third and the fourth tests because the ban was too extensive: it banned all kinds of promotional advertising regardless of their impact on energy consumption, including on electrical tools and devices that might reduce energy consumption. In short, the ban regulation was too broad and the suppression of commercial speech might prevent the exploration of associative possibilities and future developments in the form of innovative energy consumption devices and gadgets. As the Court clarified in a subsequent case, the fourth test requires the restriction should be reasonable and good enough to serve the purpose rather than be absolute.[13]

The First Amendment requires that the government formulate its restrictions on commercial speech that are narrowly tailored to advance a clearly defined government interest and that there should be a reasonable fit between the two, the means and the ends. The Court, nevertheless, affirmed in several subsequent cases that commercial speech does not enjoy the same protection as political speech. For example, false political speech is protected under the First Amendment vide *New York Times v. Sullivan* (1964), but deceptive and misleading product advertisement is not protected even if it is cleverly tied up with political and informative messages. Short of deceptiveness, commercial speech is free to expand consumers' choices through competition, innovation, hyperbole, and showmanship, of which Donald Trump's bravado, as we shall discuss, is an exemplar.

LANGUAGE AND INNOVATION

In his book, *The Shocking History of Advertising*, E. S. Turner, author and contributing journalist for the British satirical magazine *Punch*, unwittingly summed up the aspirational and innovative power of commercial speech. Advertisement, he wrote, has the capacity to conjure up the vision that "reproaches man for the paucity of his desires" and "serves as the whip" that "hustles humanity up the road to the Better Mousetrap."[14] Our dire needs and the struggle for survival do instigate innovative behavior, but it is the desire for the unattainable that fires up imagination and removes mental blocks that hinder creativity.

Hustlers, peddlers, braggarts, and showmen, however annoying at times, stir up our imagination. Ads can change the way we think about society and ourselves even though the impact may be temporary. Consider Uncle Sam's iconic image, commanding, "I Want You"; De Beers' "A Diamond Is Forever" that changed romance from song and dance and roses to something everlasting, apart from inspiring a James Bond movie; the U.S. Department of Transportation/NHTSA's "Friends Don't Let Friends Drive Drunk," whose impact has been immeasurable; Apple's "Think Different" series that transformed geeks into game changers and social revolutionaries; Lyndon B. Johnson's "Daisy" negative political commercial, which for better or worse, transformed political campaigning in the United States; the *New York Times*'s full-page civil rights advertisement "Heed Their Voices" that led to the Supreme Court landmark decision *New York Times v. Sullivan*, which established a new standard, a much higher threshold, for defamation and which freed the media from the fear of self-censorship, broadened freedom of the press, and gave us a new legal phrase *actual malice*; and most of all, consider how the simple slogan, Nike's "Just do it," and the logo Swoosh have transformed an outback sports apparel and sneaker company into a global giant, a company that has changed the way we look at the human body, the way ancient Greeks, Athenians, and Spartans used to look at themselves. Language, freed from fear, is the greatest source of innovation.

Fired Up and Get Going

Let's reflect on the television prime time of January 2004, the make-believe world of reality television. Donald Trump, the flamboyant billionaire of New York, builder of the playgrounds for the rich and famous, casinos, golf courses, and gleaming towers, and the host of *The Apprentice*, thundered:

"You're fired." The American television audience got a massive shot of adrenaline.

It is rather ironic that what we cannot say with impunity in real life we can do so with gusto on a well-scripted but supposedly unscripted reality television show. That's what dreams of the American free market capitalism are made of. The power to hire and to fire keeps people on their toes and is the quintessential American capitalist ethos and the source of its perpetual capacity for innovation and self-renewal. But civilized behavior, the desire to avoid ugliness, and perhaps the fear of legal repercussions prevent bosses from telling their underlings, "You are fired." So companies devise roundabout ways of saying the same: "We're downsizing; we're offshoring; but we're confident that a person of your abilities and talents would get a suitable position, and if you need any help, we are ready to assist you to find placement elsewhere." A required notice of separation is substituted with a severance salary, and the person is asked to clear the desk promptly. And before the evening is out, you say your good-byes and you are forgotten.

When Trump told one of the two finalists, Bill Rancic, a thirty-two-year old Chicago entrepreneur, "You're hired," waves of jubilation surged throughout the audience, more than twenty million, one of the highest of the 2004 television season.[15] As an aside, one might say that the success of the first season of *The Apprentice* showed that good television that draws huge audiences could be made without unnecessary sex and violence. Throwing gratuitous sex and violence into television is laziness. Innovation needs imagination and courage.

A superb showman and self-promoter, but someone who had trouble in straightening his own company's finances (Trump Hotels & Casino Resorts had some trouble with the U.S. Securities and Exchange Commission, and the price of a stock share hit rock bottom in 2004), "the Donald," as one of his previous wives, Ivana Trump, a Czech native, used to call him, showed the audience two of his major multimillion projects—one in California, a golf and residential complex for the rich and notorious, and the other a proposed $700 million, ninety-story gleaming tower in Chicago.[16] And then Trump asked the newly chosen apprentice to make his choice in two minutes as to which company he would like to head. Rancic promptly chose Chicago for a salary of $250,000.

It was a most spectacular materialization of the myth of the aspirational society. Trump reprimanded the loser, Kwame Jackson, a twenty-nine-year old Harvard MBA, for choosing his lead person wrongly and not having the quickness to fire the querulous and feisty Omarosa Manigault-Stallworth, who let him down with her incompetent handling of the final project, the hosting of a concert by a pop star, Jessica Simpson. But Jackson, a calm and deliberate young man, apparently did not show any regrets for the choice of his work companion Omarosa, nor did he seem to be so disconsolate for not having been finally selected as the apprentice.

The thirteen-week exposure on national television, each week ending with the firing of one of the contestants and new assignments for others, was in itself a great reward for the participants. TV exposure means millions eventually, if one knows how to capitalize on it. Even failure pays on television because it opens associative possibilities. Jackson, too, was offered a position at KFC and in the long run, probably, it would have been better for him working for a global-brand fast food company than for an unpredictable business mogul and political adventurer. Learning from Donald Trump the art of self-promotion, Jackson said that he might start a communications company to produce video games and DVD releases, among other things. "I'm an unadulterated capitalist," he told Leigh Dyer of Knight Ridder. Asked what kind of future he saw for himself in five years or so, he almost paraphrased the Persian poet-mathematician-philosopher Omar Khayyam: "On a beach, with a drink, with a Swiss bank account, a beautiful woman next to me, and a company that's been sold, and just being happy. Maybe I'll be a yoga instructor or something." Jackson was no loser. Mark Cuban, another billionaire, the owner of the basketball team the Dallas Mavericks, offered him a lucrative position to manage one of his portfolios.[17]

There's something else about *The Apprentice* that U.S. business schools need to pay attention to, since the show was watched across campuses and discussed in classrooms. Rancic and Jackson, two most remarkable young men of the post–baby boom generation, rose to the top from a hotbed of treachery, deceit, cattiness, trash-talking, and backstabbing, the cauldron that was stirred week after week for great dramatic effect to lure the television audience.

In spite of the fact that it seemed like a dog-eat-dog world of ruthless capitalism, for young aspiring entrepreneurs the lesson seemed clear that Rancic and Jackson rose to the top because they did their best to work with others and earned their respect and cooperation. They were innovative, bold, and courageous, the kinds of qualities in successful business leaders that business schools try to impart to their students. It is with this young, rising generation of leaders in mind that corporate America is confident about itself, though political America shivers in its made-in-China shoes when it sees Chinese goods flooding the global marketplace. China presents very little threat to America.[18]

The threat to America comes from within, for example, the lack of transparency on Wall Street. Neither the marketplace nor big business listens to people on Capitol Hill, or our man in the White House. It is imperative, therefore, that the news media keep a hawk-eye not only on the world of corporate finance and high-stakes wheeling-dealing but also on chicken thighs and women's legs.

Beautiful Deceptions

Barring some celebrities and entertainers, very few people ever get invited to the Victoria's Secret Fashion Show parading its angels to promote a most seductive brand of lingerie and sleepwear in a high-profile setting of glitz and glitter. Most people console themselves by watching the show on television, and even at the risk of being told that they might have sour-grape syndrome, they feel that the show is nothing but soft pornography.

Television entertainment programs invariably focus on the voyeuristic and the lascivious in order to keep the audience glued to the screen so that they don't escape commercials. Television as a business platform does purvey some serious culture, but mostly it is pop culture, the culture of the marketplace, where doing transactions through various inducements and subliminal persuasion is the dominant goal.[19]

A curious thought crossed my mind once upon a time while I watched a Victoria's Secret show. What would happen to these gossamer, curvaceous, bony models with wings of angels walking on cushions of air if they were fed for a year on a "wholesome" diet of KFC? Would there be any Victoria's Secret left? This perhaps sums up the modern American tragi-comedy: pig out on whoppers, fried chickens, and sundaes, and then—lured by Victoria's Secret—pop in diet pills to slim down fast. Some have their protruding fat surgically scooped out, others go to quick-fix lawyers and sue fast food chains for making them obese. America is a beautiful, litigious country. My hometown lawyer crows all the time: "Public safety is my first obsession, but if you get hurt, come to me, I am better than the best and the rest."

Obesity, which has reached an epidemic proportion in the United States, can't be controlled with diet pills or genomic medicine, wrote Dr. David L. Katz of Yale Medical School in the *Wall Street Journal*. Only a major cultural and behavioral shift, away from vending machines, fast food, and video games, would take us to the never-never land where men are all muscles and women are all Victoria slims. "Epidemic obesity," according to Katz, "is arguably the gravest public health threat in the United States today. Current trends, should they continue, will see heart disease in adolescents; more deaths related to diet than to tobacco, drugs and alcohol combined; and a generation of adults with a shorter life expectancy than their parents."[20]

Entombed between KFC and Victoria's Secret, the United States has created a multi-billion-dollar quick-fix diet and weight reduction industry. Since environmental and behavioral changes can't be socially engineered in a market-driven society, what is the solution? A watchful press and receptive courts, of course, are categorical imperatives. Suing those who make misrepresentations about health benefit claims for their fast food draws the attention of the news media and the regulators.

Deceptiveness, of course, comes in many forms. Tobacco companies have learned the lesson, especially since Jeffrey S. Wigand, a former vice president of research and development at Brown & Williamson who turned whistleblower, told CBS's *60 Minutes* how the industry was creating smokers' addiction by manipulating nicotine content and conducting massive campaigns to conceal the health dangers of smoking that causes lung cancer.

When a KFC's advertisement called its fried chicken a health food that consumers could eat guilt free, without worrying about weight gain, it drew widespread protests from consumer advocacy groups and the media. In two separate advertisements that subsequently were pulled, KFC claimed that its chicken breasts were less fatty than Burger King's whopper and furthermore that its original recipe chicken would fit into a low-carbohydrate diet regimen. In one advertisement, a youthful dude asked another one in surprise: "Is that you? Man, you look fantastic! What the heck you been doin'?" Speaking through his stuffed mouth, the other man replied, "Eatin' chicken." And the voice-over bragged: "So if you are watching carbs and going high protein, go KFC." Then the punch line: "For a fresh way to eat better, you've gotta KFC what's cookin'!" In another advertisement, a woman plopped a bucket of KFC original recipe chicken before her male friend's awestruck face and said, "Healthy eating 'starts today.'" Of course, there were declaimers such as "not a low fat, low sodium, low cholesterol food," but those who want to hog and also keep their weight down do not pay attention to fine print. The ads appeared to be some of the worst deceptive advertisements at a time when public health authorities are extremely concerned about the epidemic of obesity. The Center for Science in the Public Interest filed a complaint with the Federal Trade Commission that KFC's claim that "fried chicken can, in fact, be part of a healthy diet" can't be substantiated. It's nothing short of ridiculous and extremely misleading.[21]

In this permissive culture of eat-all-you-can-eat-and-lose-weight diet hypes (and discover Victoria's Secret), the Federal Trade Commission has reasons to remain hawk-eyed about deceptive advertisements. FTC rules require that an advertisement's objective claims must be substantiated with scientific research. In case a claim can't be substantiated, FTC asks the company to issue a corrective ad for a specific period announcing that its previous claim was false. Warner-Lambert Company, for example, was asked in 1972 to issue a corrective ad, stating: "Contrary to prior advertisement, Listerine will not help prevent colds or sore throat or lessen their severity."[22]

While FTC and the appeal process take years to decide a case, in a market-based democratic society, a model that is being copied in many developing countries, it is the responsibility of the media to protect the consumer. Consider an *Ad Age* editorial reprimand: "In the long history of absurd, misleading and ludicrous claims," said the editorial, "the campaign's positioning of KFC's breaded, fried chicken as a part of a healthy diet merits special derision. It damages the credibility not just of KFC but of the marketing industry."[23]

But that still leaves one totally wondering that if Victoria's Secret angels don't eat KFC fried chicken, what do they eat that makes them walk like fairies in Paradise? Here's the dilemma: While you can fight KFC, you can't fight Victoria's Secret. Victoria's Secret may be the most beautiful deception on earth, so buyers beware when business schools teach their students: Brand is the thing, and that is all you need to know. Without a free and fearless press, branding is nothing but deception.

THE CULTURE OF ASPIRATION AND THE ARAB STREET

One of the most widely practiced forms of indoctrination through seduction, advertisement creates and transforms our desires and wants into compelling needs, a fascinating field of study that has been unfortunately ignored by soft power advocates and counterterrorism experts.

Just as the culture of fast food has been driving people to McDonald's, Pizza Hut, and KFC, for example, to consume billions of chickens, pies, and hamburgers, and tons of French fries, the culture of Paradise, the afterlife to be spent in the company of beautiful women that Islamic militants, Al-Qaeda, and their decentralized networked franchisees promise to the Muslim youth, has been driving them to jihad and suicide. Both cultures unabashedly promise fulfillment: here or hereafter. Communism promised earthly seductions, a stateless paradise of equality. As Vladimir Lenin said, "The state will be able to wither away completely when society adopts the rule: 'From each according to his ability, to each according to his needs,' i.e., when people have become so accustomed to observing the fundamental rules of social intercourse and when their labor has become so productive that they will voluntarily work according to their ability."[24] Yes, indeed, the Soviet Union did wither away because it could not develop the culture of aspiration.

In the final analysis, the subliminal seductions of the Islamic militant jihad are no different from what communism had to offer in the intoxicating and revolutionary days of Leninism and Maoism. Communism, however, could not withstand the soft and seductive power of free market capitalism best expressed through consumer culture and its endless capacity for self-renewal. Militant Islam and jihad, too, will eventually dissipate due to the constant and relentless confrontation with the persuasive power of consumer culture that embraces women and puts them on a pedestal of desire: "You want to know what comes between me and my Calvins? Nothing."[25]

Al-Qaeda and militant Islamists have been creating the same psychological conditions under which a person's desires and dreams become compelling needs. According to psychologist Abraham Maslow, there is a hierarchy of five primal needs that drive human beings to action to seek satisfaction: physiological, safety, social, esteem, and self-actualization.[26] A person tends to fulfill his physiological needs for food and shelter first before he seeks to satisfy other higher-order needs. But experience shows that once the basic needs of food and shelter are satisfied, some people seek to satisfy other needs simultaneously instead of seeking their satisfaction one after the other in a hierarchical order. In fact, the need for self-actualization, realizing one's potential, summed up best in the U.S. Army slogan "Be all you can be," may become so overpowering in a person due to indoctrination that he may minimize or even forego other needs.

Industrialization, mass media, and advertisement have turned the United States into a culture of aspiration and of high consumption, although it wasn't always so. Beginning with the Penny Press in early nineteenth century and later on through the yellow journalism of William Randolph Hearst, news media began to depend increasingly on advertisement revenues. Business and industry needed advertising to reach masses in order to increase sales. Thus began the symbiotic relations among the three—the media, advertising, and industry—to create the American consumer whose desires must be cherished and wants must be transformed into dire needs.

By turning malls into places of leisure and pleasure, marketers and advertisers have been transforming shopping into an enjoyable experience. Metaphorically speaking, "Shop 'til you drop" pursuits leave little time for any thought for the afterlife.[27] The day after Thanksgiving, Black Friday, is a unique American experience, the epitome of the aspirational society. Christmas has become a secular marketing experience and a major driving force for the economy. So will become Ramadan one day as the Arab Street driven by global mass media and the Internet inevitably becomes a free market capitalist consumer society. Like Billy Graham, Jerry Falwell, and other Christian fundamentalists, the Mullahs and the Islamists, too, will realize that the pursuit of happiness in the Arab Street is more realistic and important than the dreams of Paradise.

By bringing consumers into a desirable media mix, by segmenting the population into demographics and psychographics, and by demanding that media companies create a cultural program that not only supports commercial products but also creates a cascade of compelling needs, advertisers have created a culture of desire that makes people work harder so that they can buy and consume more. Every year, for example, the auto industry comes up with new models with varied psychological appeals and lucrative incentives for the consumer to get rid of his old car even if it is in good condition and go for a newer model. The advertising industry created a middle-class suburban woman, the "soccer mom," and told her that she needed a van to chauffeur her children from school to the ballpark.[28] Ad men have made SUVs a vehicle of choice for many Americans. After 9/11, auto manufacturers saw the consumer need for greater security, and today's SUVs give a feeling of strength and power like armored vehicles.

Advertisement is the best practical application of what Aristotle said about rhetoric: "The faculty of observing, in any given case, the available means of persuasion."[29] Advertisers have been using target marketing in order to reach and persuade their audiences more effectively. Similarly, instead of looking at the Muslim population of a country as a monolithic mass, it should be segmented demographically for specific messages. For example, the message for a Muslim woman with her overpowering needs for personal security, family, and children has to be different from what is aimed at the Arab youth.

It is not too farfetched to theorize that militant Islam, like communism, will dissipate due to the aspiration of the consumer society with its promise of happiness through instant gratification rather than the glory of the virgins of Paradise that the militant jihad promises through sudden-death experience. The war against terrorism has to be fought in the battlefield of ideas, by creating aspirations, by creating desires, wants, and needs that can be fulfilled. Al-Jazeera and Al-Arabia, West Asia satellite networks that did not hesitate to broadcast messages from Osama bin Laden, out of fear or favor, would have no problem in airing commercials that persuade Muslims to reach for their wallets rather than their guns. Nike, Coca-Cola, McDonald's, Hollywood, and hip-hop, figuratively speaking, will ultimately defeat the destructive culture of jihad.

Militant Islam, with its bloody mission of worldwide jihad, will go the way of Soviet communism, with its bloody global revolution. The Arabs would rather trade with the enemy than kill him, if history has any lesson. Spreading consumer culture, not misguided YouTube videos such as "Innocence of Muslims," is the way to subvert militant Islam and transform the Arab society. The Arab uprising that began in Tunisia and Egypt and seared through Libya and Syria will ultimately flower into the munificence of a secular, free market, democratic culture, even though the Muslim Brotherhood might stage a temporary claim to power. Muslim festivals such as Eid al-Fitr will become as secular as Thanksgiving, a holiday season for a shopping bonanza that celebrates the triumph of the aspirational society.

TURNING IDEAS INTO ASSETS

The United States is desperately in search of ideas that can make things happen, whether it is to spot Al-Qaeda operatives to drone them out of existence, or to persuade the reluctant shopper to take out the credit card and spend whether she has the money or not in the bank. The day after Thanksgiving through Christmas is by far the busiest shopping season in the United States. A drop of even 1 percent in holiday sales ripples through the marketplace and eventually every home and every trailer park. To forestall the economic disaster, shoppers are offered unprecedented discounts on sales of all kinds of goods from cars to carpets. Any creative idea that brings the shopper to the mall and persuades her to fill up the shopping cart is an invaluable asset.

But how do you transform an idea into a marketable innovation? "I am your idea," said an Accenture blurb. "One day you'll look for me and I'll be gone."[30] Ideas are transient unless you grab them, change them into design, and make them do something practical. Make ideas work by sharing with people who know how to turn them into innovations and tangible goods, the Accenture ad said. China's sustained economic growth of 8 to 10 percent during the last three decades comes up for scrutiny and envy every now and then.

The legend goes that one day the Chinese supreme leader Deng Xiaoping had a revelation: Capitalism is good, he muttered after returning to China from a visit to the United States. Work and make money, he said, and the state-controlled sluices of entrepreneurial spirit opened up in China. Economic progress without political freedom is regarded as the Chinese miracle, which would not have happened if the United States had not opened its markets and let the capital flow to the communist nation. China, according to Keith Bradsher of the *New York Times*, "by quickly converting much of its economy to an unfettered and even rapacious version of capitalism, has surged far ahead. . . . China has high-speed freeways, modern airports and highly efficient ports that are helping it dominate a growing number of manufacturing industries."[31]

In a matter of years, China has become the manufacturing hub of the world, lapping most foreign direct investments. All ships are sailing to and fro to China. China's economic miracle is not founded on any grandiose economic theory, but on a few simple ideas: good-enough law-and-order conditions without the rule of law; good and reliable transportation and communications facilities; and the confidence to let the people go and make money without letting them question the Communist Party rule. China borrowed communism from the Soviet Union and formed a strong, unified, authoritarian state, which today is managed like a giant corporation under an elite, nationalistic collective leadership. China borrows, assimilates, and keeps growing. The American genius is different.

Mashing Up

How to take ideas from one field and make them work in another is what Americans are good at, sometimes brilliantly so, which exemplifies the concept of adjacent possibilities. For example, American advertisers have used Jean Piaget's theory of child development, sensory experiences, and visual stimulation to sell EZ Squirt ketchup to grown-ups. Alissa Quart wrote in *Wired* magazine, "Piaget is only the beginning. Just as the pharmaceutical industry steers medical research, marketing and advertising are beginning to guide the way scholars investigate brain functions, perception, and language."[32]

Cognitive science, a multidisciplinary area that includes psychology, neuroscience, sociology, and computer science, is associated with the study of artificial intelligence and autonomous systems, but now its ideas are being increasingly used to study "the psychology of acquisition and the science of material desire" for better marketing and placement of products—anything from toys and cereals to jeans. Some of us do have qualms about turning academia into a handmaid of the marketplace, but in America today walls are breaking down and various fields of intellectual endeavor are mixing and mingling, creating new possibilities.

Ideas flow from one field to another and flourish wherever they find the best applications, whether it is the shopping cart or fighting terrorism. In the ultimate analysis, it is all about the psychology of desire that transforms an idea into an asset, an idea that turns driving a car into love and adventure; jihad into a suicidal desire for the virgins of Paradise; and China to become a global manufacturing hub from toys to smart phones, of course, designed by corporate America. All this mash up from contrarian fields of ideas makes advertisement a mighty force for innovation, which could be enhanced or stymied by patents and intellectual property laws.

BUSINESS METHODS PATENTING

Lately companies have been rushing to patent their unique methods of doing business, thus creating impenetrable walls to hide ideas that should be tested and debated in public. Vivius, Inc., a Minneapolis-based company, claimed a business method patent by calling itself "the architect of a core technology and innovative transaction model that allows consumers to calculate a unique insurance premium for each participating family member, based on their individual preferences and choices of co-payment levels and providers."[33]

Vivius offered its Internet-based model "to consumers through insurance carriers under a turnkey private-labeling agreement." Once an activity is called a business method or a technology, the owner could ask for a patent. Ridiculous though it may sound, even one's tumbling walk, swagger, or a Dirty Harry look could be called a method of doing business and be patented. The phrase *turnkey private-labeling agreement* is new, but what is unique about it? If privacy can be protected through any unique business method,

Vivius should have exhibited it to the public and got it tested in the marketplace. The fact is that as soon as a person logs on to a website, he leaks all personal information, in spite of his best intentions. A promise of total web privacy in doing business is questionable. That said, there is indeed an urgent need for developing a system that allows a person to shop for the best and least expensive clinics, hospitals, and drug stores wherever they are located in the United States, or elsewhere. The cost of drugs and medical malpractice insurance is ripping the health care system, for which the answer is Obamacare, not the Vivius transaction model, because it does nothing to ameliorate the condition. Until lawmakers sort out how to regulate the growing field of e-commerce, Congress should put a moratorium on patenting Internet business methods.[34]

Every system in the course of time tends to dissipate and become corrupt unless it is renewed. And that is now happening with the system of granting patents, which Congress, under Article 1, Section 8, established to "promote the progress of science and useful arts, by securing for limited times to authors and inventors the exclusive rights to their respective writings and discoveries." But in this swirling chaos of greed and more corporate greed that is threatening innovation and global e-commerce, it is time for Congress to act. The cessation on granting e-patents will allow entrepreneurs to use each other's business methods without any fear of infringement of intellectual property rights, which will serve the societal purpose of expanding e-commerce better than granting an Internet patent for every clever stroke of doing business in a slightly different way.[35] If the constitutional objective of encouraging innovation and creativity can be accomplished in a less damaging way that at the same time maximizes societal interest, it must be tested. Unfortunately, patents have become a form of corporate aggrandizement, as has happened in the *Apple v. Samsung* case.[36]

There is a historical precedent of dealing with this kind of muddled situation, as it happened in the early days of the development of radio. During World War I, radio was beginning to develop into a medium of promoting business and long-distance shipping and military communication. There were many contentious patent disputes between Marconi Wireless Telegraph Company and other inventors. In order to put a stop to patent disputes and advance radio's use under U.S. control, the U.S. Navy was given monopoly over radio technology patents until the end of World War I. Pooling patents hastened the development of radio, which became a dominant cultural and business medium after World War I. Internet and e-commerce patenting today is in a similar stage of development.

The current system of granting business method patents began in 1998 with *State Street Bank & Trust Co. v. Signature Financial Group* regarding the Hub-and-Spoke mutual fund system.[37] It surged with Priceline.com "Name your price," and it became a hot-button issue when Amazon.com claimed victory over Barnesandnoble.com regarding the patent on the "1-Click" business method—a development that was nothing but a negation and perversion of the original purpose of the patent law. On July 7, 2011, the European Union Patent Office denied Amazon the patent claim on 1-Click, ruling that the claim lacks an inventive step.

By putting a limited moratorium over Internet business method patents, Congress will help e-commerce develop rapidly and encourage inventiveness and creativity. It is time to balance the public interest with the interests of big business. The public interest in innovation will be better served by raising the standard of "nonobviousness and novelty" to never-before-experienced extreme uniqueness, which should, of course, be open to public discussion in the media.

A variation on an old theme or a method of doing business is not uniqueness deserving a patent. If a business method is not 95 percent new and nonobvious, it does not deserve any patent. Anything less than the "95 percent new and nonobvious" standard should be regarded in the realm of prior technology, and it should belong to the public domain. The argument that businesses won't invest money in developing new business methods is groundless. The momentum of e-commerce is so powerful today that nothing could stop investors from investing money into new e-commerce ventures.

E-commerce would have risen manifold if "Name your price" and "1-Click" were placed in the public domain rather than given proprietary control to already well-established businesses. Amazon.com and Priceline.com, as successful business concepts, need to be further developed, and that can be done only through the challenge of the marketplace. In the absence of any initiative by Congress, it is left to the Supreme Court to subject business methods claims to strict scrutiny, keeping in mind that the ultimate goal of granting patents is to encourage innovation.

In the case of *Bilski et al. v. Kappos*, in which plaintiffs had sought a patent on a procedure to hedge the seasonal risks of buying energy, the Supreme Court ruled that the process at issue could not be patented; nonetheless, they kept it open that some business methods could be patented.[38] That's the way to go about Apple's patent claims about its "borrowed" uniqueness. But if the courts cannot correct the aggrandizing behavior of powerful corporations, unfettered and robust media must do.

CORPORATE SPEECH

In 2010 the U.S. Supreme Court ruled in *Citizens United v. FEC* that the First Amendment protects corporations' right to political speech, and corporations may use general treasury funds to make independent expenditure in supporting or opposing political candidates during elections.

The decision has aroused tremendous fear that corporate speech will drown out genuine democratic voices and result in corrupting political processes.[39] The decision is also recognition of the fact that the exercise of political speech, in fact any form of public speech, without money, in the age of mass and social media, is impossible. To be heard, individual voices have to be aggregated and organized, for which money is indispensable. How much money is good enough, and what should be the sources of the money has been a continuous thematic thread and motivation for all campaign finance reforms since *Buckley v. Valeo* in 1976, in which the Supreme Court upheld limits on individual contributions but struck down limits on campaign expenditures, independent expenditures by individuals and groups, and expenditures by a candidate from personal funds. The disclosure and reporting requirements and the public financing scheme of the 1971 Federal Election Act were also upheld.

The Bipartisan Campaign Reform Act of 2002 (BCRA), the so-called McCain-Feingold Act, was aimed at eliminating the role of soft money in campaign financing. This was done by barring national political party committees from raising or spending any funds not subject to federal limits. The second major focus of the Act was to prohibit "electioneering communications" in the form of broadcast commercials that identify a federal candidate within thirty days of a primary or caucus or sixty days of a general election. The Act also prohibited any commercial paid for by a corporation or paid for by an unincorporated entity using corporate or union general treasury funds. The Supreme Court decision in *Citizens United v. Federal Election Commission* overturns this key provision of the McCain-Feingold Act.

Whether the unleashing of unlimited money by corporations, unions, and advocacy groups into the political and electoral processes will corrupt democracy or not will depend upon the role the press, social media, whistleblowers, and shareholders and corporate board members play in demanding transparency. Just as speech can neutralize speech, money can also neutralize money. In the 2012 elections, it was PACs against PACs, Super PACs against Super PACs. Corporate money played little role in the reelection of Barack Obama or the defeat of Mitt Romney.

In the long run, *Citizens United* might prove that money cannot buy elections, although it might make media companies rich by pumping billions of dollars via campaign ads. The ruling will further open corporations to strict media scrutiny and greater transparency as to which political and social causes they believe in and which political candidate for electoral offices they support.

Corporate stakeholders, nongovernmental organizations (NGOs), and special interest groups will force corporations to explain their polices and political behavior more rigorously than they have been doing so far. Corporate boards under pressure from the public and the press will be compelled to become more transparent as to why their advocacy of certain political issues will serve the corporate interest. Various manifestations of the Occupy Movement will compel corporate America not to ignore the public interest and their social responsibilities. Congress may be compelled to legislate corrective measures to stop abuses. *Citizens United* recognizes that money is not evil, that corruption can be fought through exposure, which is the prime responsibility of the press, social media, whistleblowers, and organizations like WikiLeaks.

Chapter Five

Free Press and Marketplace Fundamentalism

THE DARK SIDE OF THE ASPIRATIONAL SOCIETY

Marketplace fundamentalism, ruthless capitalism, leaves all problems to be solved by the market forces of supply and demand and has been credited with fueling unprecedented economic growth in the United States. But frequently it also leads to corporate scandals, triggering the loss of faith in the marketplace. It is not only a few bad apples—it seems that sometimes the whole crop is rotten. To pacify the sullen, angry public, corporate scandal sheets are exposed and miscreants handcuffed. The dominant public sentiment, especially when a big corporate scandal hits the headlines, persists that leaving everything to the marketplace is dangerous, and more oversight is needed.

Today many countries, following the American economic model, are committed to the divestiture of state enterprises and privatization to speed up economic growth and to create jobs. But privatization requires the restoration of faith in the fairness and efficiency of the market. It is necessary to improve upon the American model of free market capitalism so that other countries can follow it without undergoing the trauma of the bubble and bust that the American people have been experiencing periodically.

The marketplace must serve everyone's needs and become a tool of spreading wealth to the middle class and the poor. It is possible that the psychology of diffused endless wars, homeland security, and terrorism have made Americans more insecure. Instead of greater freedom and deregulation, which free marketplace capitalism needs for economic growth and creation of wealth, there is a demand for greater governmental authority in public life and stronger oversight over corporate governance. So in these strange times, the emerging wisdom is: control capitalists to save free market capitalism and control freedom to save political freedom. But can you nurture the free market and encourage economic growth with hands-on controls? Too many controls choke the free market. On the other hand, without controls you let loose in the marketplace greedy predators like Gordon Gekko, who said:

> The richest one percent of this country owns half our country's wealth, five trillion dollars. One third of that comes from hard work, two thirds comes from inheritance, interest on interest accumulating to widows and idiot sons and what I do, stock and real estate speculation. It's bullshit. You got ninety percent of the American public out there with little or no net worth. I create nothing. I own. We make the rules, pal. The news, war, peace, famine, upheaval, the price per paper clip. We pick that rabbit out of the hat while everybody sits out there wondering how the hell we did it. Now you're not naïve enough to think we're living in a democracy, are you buddy? It's the free market. And you're a part of it. You've got that killer instinct. Stick around pal, I've still got a lot to teach you.[1]

This is the dilemma of cutthroat capitalism. In a meeting with corporate crime fighters in 2002, President George W. Bush said, "Your mandate is to root out crime from corporate America's boardrooms and executive suites."[2] Bush portrayed himself as being as tough on corporate crimes as he was on terrorism.

In the Bush administration era, the media show-and-tell tales about corporate crimes and terrorism gave the impression that both had the same origin. While terrorism is an international problem and needs worldwide cooperation to eliminate the scourge, corporate crimes can be fought and eradicated through legal and structural reforms that mandate transparency in corporate governance, supported by a vigilant press and social media that are always on the lookout for corporate malfeasance.

SOURCES OF CORRUPTION

There are three major sources of corruption that have bedeviled corporate America. First is the problem of the initial public offerings (IPOs) of hot stocks, which Wall Street investment banking firms offer to company executives as a quid pro quo in the hope of getting financing business from them.[3] Many CEOs have made millions by selling their IPOs the same day when the stock was issued and prices went up, in some cases several hundred percent of their initial value. By separating IPO allocation from investment banking, the conflict of interest can be eliminated; therefore, as a first step toward accountability, company executives should be required to disclose the IPO shares they receive from investment banking firms.[4]

The second source of corruption originates from market research analysts of investment banking firms, on whose recommendation investors depend for buying and selling shares. It is rarely that a market research analyst of an investment-banking firm will be bold enough to tell investors the truth that the stock of a company, the client of his firm, is going down even if it is so.[5] He might lose the job. So he says, buy or hold, even when he should be advising investors to sell.

The third problem occurs when accounting companies, who should work as public watchdogs, compromise their integrity by offering consulting services to the same clients whose accounts they audit. The accounting firm of Anderson Consulting, for example, instead of being a watchdog of Enron, became its lap dog and was done in along with its master.[6]

Not long ago, Milton Friedman, Nobel Laureate in economics, asserted that marketization, deregulation, and privatization are the keys to economic growth and prosperity.[7] But Friedman's free market equation needs another element; that is, democratization and accountability of power. Privatization without democratization leads to corruption, especially when major corporate interests also control the media. But the situation need not be totally hopeless because it is not possible to buy all journalists in a free society. Aymo Brunetti and Beatrice Weder, rather sanguinely, conclude that

> the more involved a corrupt arrangement the more fame an investigative journalist can earn by uncovering it. As long as there is free entry in journalism and in publishing—which is one of the defining features of a free press—it will be difficult to form an effective cartel which encompasses all journalists. . . . The reason is, that journalists have incentives to uncover corruption and it is unlikely that all journalists can be brought to cooperate in corrupt arrangements. Such a cartel would be difficult to sustain in an environment with many independent journalists and high reputational profits from uncovering corrupt arrangements.[8]

Corruption and incompetence in journalism is a serious problem, which nonetheless can be handled legally and professionally. Of much greater concern is the fact that today not enough journalists are being trained in complicated areas such as finance, banking, corporate governance, technology, and medicine. When a disaster occurs, senior reporters and editors promptly call upon their panel of experts to explain what happened, whether it is nuclear disaster in Japan or financial meltdown on Wall Street. But most of them, even journalists in the financial media, do not have deep knowledge and interest in these highly complicated areas required to do investigative reporting so that society can take preventive measures.

On the anniversary of the collapse of Lehman Brothers, Cristina Romero of the European Journalism Center asked poignantly as to "what was happening in this media space before the economic breakup? What was the focus of financial journalists? Did a glittering market blind them all? Above all, has financial journalism played the role it was expected to?"[9] Financial journalists are not only poorly educated but they are also in short supply, which is also the case in other highly specialized areas. Corporate corruption has a high correlation with media ignorance.

George W. Bush, in spite of being probusiness, used to speak forcefully, and in fact, with the passage of time he became increasingly more eloquent about corporate malfeasance, including inside trading, creative accounting, misuse of stock options, and tax evasion by moving companies' registered offices to tax-sheltered offshore islands. He proposed tough measures including longer jail terms for corporate executives who defraud their companies, employees, and shareholders. He proposed a financial SWAT team, a quick-action corporate task force to catch crooks. He asked that chief executives vouchsafe the accuracy and veracity of the financial statements. He recommended that the accounting industry be supervised by an independent board, and much more. But corporate America paid him no heed. Wall Street had heard it all before and nodded in silent indifference. Like George W. Bush, Barack Obama, too, the leader of the greatest economic and military power in the world, has had little effect upon Wall Street, which has been behaving like an independent sovereign body responsive to its own internal dynamism.

The only time one sees the market being affected in its up-and-down movements is when the Federal Reserve Bank changes its rates, there's some danger to the world oil supply, or there's some other financial market hiccup. Wall Street is a strange, unpredictable, monstrous manmade system. Nonetheless, millions of Americans and many millions more from the rest of the world depend upon it, have no choice but to trust it, and perhaps have helplessly invested in it. Since Wall Street is nothing but an epitome and projection of how Corporate America is working or what's wrong with it, it should worry politicians because the financial interest of millions of people, including their political constituencies, is involved in it. Besides, Corporate America feeds their political careers by financing their election campaigns, and there is always the next election to win.

American politicians have an insatiable appetite for campaign money, and only Big Business can quench it. As a quid pro quo, politicians relax regulatory oversight either through deregulation or by providing insufficient funding to regulatory agencies. A regulatory agency starved of funds cannot hire enough staff to watchdog corporations, and that is the payback time for Big Business.

Jack Abramoff, a most notorious lobbyist and corrupter of American politicians during the times of George W. Bush, was at the center of one of America's biggest corruption scandals. Abramoff spent more than three years in prison for his crimes. Now a reformed man, Abramoff discussed with Lesley Stahl of CBS his modus operandi of influencing and buying Washington politicians and their aides through sumptuous job offers and lavish gifts. Nothing has changed in spite of the reforms Congress instituted. Here is an excerpt from the conversation:

> Stahl: Was buying favors from lawmakers easy?
> Abramoff: I think people are under the impression that the corruption only involves somebody handing over a check and getting a favor. And that's not the case. The corruption, the bribery, call it, because ultimately that's what it is. That's what the whole system is.
> Stahl: The whole system's bribery?
> Abramoff: In my view. I'm talking about giving a gift to somebody who makes a decision on behalf of the public. At the end of the day, that's really what bribery is. But it is done everyday and it is still being done. The truth is there were very few members who I could even name or could think of who didn't at some level participate in that.[10]

While Obama is perceived as a member of the elite intellectual class in spite of his humble interracial origins, Bush was regarded by Corporate America as a business-friendly person. His own past as an oil executive in Texas came under close scrutiny. He might have benefited from the very loopholes that he asked Congress and Corporate America to eliminate. At a press conference on July 9, 2002, defending his supposedly smart or deviously well-timed 1990 sale of stock in the Harken Energy Corporation, Bush as a company insider blamed Harken's lawyers for his failure to report the sale in a timely manner as required by law, and said, "In the corporate world, sometimes things aren't exactly black and white when it comes to accounting procedures."[11] But they ought to be made as clear as daylight to maintain trust in the marketplace.

Accounting rules must be absolute, absolutely trustworthy. Blending black and white is what is called *creative accounting*, a euphemism for cheating and fraud. Accountants and auditors are the fiduciaries of public interest.[12] They should be as independent from Corporate America as the U.S. Supreme Court is from the executive and Congress. Many political commentators remarked that Wall Street did not respond to Bush's righteous overtures because of his own past business dealings that might have been regarded in some circles as neither black nor white, apart from his close associations with Big Business that supported his political campaign. But the market does not care for President Obama either.

Reacting to the perceived softness of President Bush toward Big Business, the Senate passed tougher measures to punish corporate misdeeds and restore public trust in the free market.[13] While signing the Sarbanes-Oxley Act, Bush said, self-righteously:

> No more easy money for corporate criminals, just hard time. . . . tricking an investor into taking a risk is theft by another name. . . . The era of low standards and false profits is over . . . No boardroom in America is above or

> beyond the law. . . . Free markets are not a jungle in which only the unscrupulous survive, or a financial free-for-all guided only by greed . . . For the sake of our free economy, those who break the law—break the rules of fairness, those who are dishonest, however wealthy or successful they may be—must pay a price.[14]

The emphasis has been more on punishment than on preventive measures. Many investors who saw their retirement future disappear wanted the CEOs of Enron, Global Crossing, Tyco, WorldCom, and other companies in orange jumpsuits and paraded before television cameras as common criminals. The 2002 congressional elections were closing in on them, and both Republicans and Democrats wanted to appear mean and tough on Corporate America. When Bernard J. Ebbers, former CEO of WorldCom, exercised his right under the Fifth Amendment refused to answer questions at a congressional hearing, he was treated by the members as a common thug and was deservedly humiliated. The Fifth Amendment protects a person from self-incrimination and enables him to maintain his innocence unless proven guilty, a common ploy used by corporate criminals when hauled before congressional committee hearings.

Wall Street controls the commanding heights of global finance. Its impact on world markets is like that of a tsunami; it is not a butterfly effect. The rest of the world is as much interested in the proper functioning of Wall Street and Corporate America as are and should be, no doubt, the American people. Many countries aspire to follow American-style marketplace democratic capitalism as a model for economic development. Privatization of industry without a reliable system of checks and balances can be disastrous for a developing country.

Accountability is the essence of marketplace democratic capitalism. What is happening in Corporate America today is not a pretty picture. Business heroes of yesterday are biting dust today. Though trust in business has not totally vanished, it is much diminished. The United States must show the way and demonstrate how to encourage entrepreneurial creativity and innovation without unleashing capitalist greed. One of the most effective ways to throttle Gordon Gekko's "greed is good" energy and redirect it for constructive social purposes is through public revelations by the news media. Greed can be exposed through transparency, which is the most important function of freedom of the press.

INSATIABLE AT THE TOP

Until 2001 when he retired from the company, General Electric's global corporate visionary Jack Welch was the toast of the corporate world. During his tenure as CEO, the total market capitalization of the company rose by

$400 billion. On the occasion of his retirement, Welch published a best-selling book, *Jack: Straight from the Gut*, which made him all the more popular on the lecture circuit, where he talked of globalization, e-business, "boundary-less" sharing of ideas, and focus on people.[15] In fact, he was so much focused on corporate success as the CEO that he fired more than one hundred thousand employees (those who couldn't deliver), and he came to be called "Neutron Jack."

The neutron bomb kills people but spares property, and no wonder the stock market loved Welch. The blurbs, book reviewers, and opinion-makers called Welch a rebel, a self-made man, a leader, a business genius, and all the good words that money could buy, until his wife of thirteen years, Jane Beasley Welch, hung the family dirty laundry in public, asking for a divorce from her husband, who was estimated to be worth more than $900 million.[16] Why would any woman walk out on a man regarded by all and sundry as a corporate superman, someone who had become a role model for the up-and-coming stragglers and strugglers on the corporate ladder? Jane Welch discovered that her sixty-six-year-old husband could not control his libido. While being interviewed by the editor of the *Harvard Business Review*—a woman twenty-three years younger than him—the retiring CEO developed a soft spot for her and eventually married her.

That is how we know the rest of the story, because divorce papers are public documents and tell more than a butler, kitchen maid, or tabloid gossip could ever tell. In the immediate post-Enron age when executive compensations were being subjected to close scrutiny, Welch's retirement benefits and perks seemed to be obscenely exorbitant. His salary of $16.7 million for the last year of his service as CEO of General Electric did not raise envious eyebrows, but what drew the attention of the public and the media were his irrevocable retirement benefits. Consider some of the retirement perks, according to media reports and based on court documents: A minimum of $86,000 annually as a GE consultation fee; most of the living expenses in the exclusive Trump Towers, where the monthly rent alone was $80,000; the use of GE's Boeing 737 business jet; and company-paid membership in some of the most exclusive country clubs and the best seats at the top sports events, apart from the annual pension of $9 million.[17] All this would have remained confidential and Welch would have continued to be a corporate hero but for the affair that broke up the marriage. Not every woman can muffle her cries and sing, "stand by your man" when he is having a good time doing things you don't understand, as Hillary Clinton did and stood by "her man."[18]

We are living in a technology-enforced transparency age. Transparency about personal life is not limited to the rich and the vulgar of American society. Everyone's private life is up for grabs in the age of the Internet. It is not only divorce cases that attract neighbors' attention; arrest warrants, bond postings, places where sex offenders live, traffic tickets, parking violations, complaints about domestic abuse, all sorts of sordid affairs, except perhaps medical and financial records, are available online now.[19] Speedy and inexpensive justice would have been a better service to the community, but now justice is becoming public flogging.

Earlier, too, most of the court documents, unless they were sealed by the judge's orders, especially those involving juveniles, were available to the public, but it took so much time and effort to put together, and only the toughest investigative reporter or gossip columnist had the audacity to do so. So there was privacy by default. What was not available was private. But thanks to the Internet, the line between what is public and what is private has been vanishing in the United States, the most advanced digital society where some people think that data shredding should be deemed as free speech protected by the First Amendment. Jack Welch could not have shredded his past selectively. One could not but sympathize with Jack Welch. Perhaps he was more sinned against than sinning, but that is the price one pays for being a public figure and being too successful in America.

In a *New Yorker* cartoon, an office worker talking to his boss says, "Good news, chief, a computer virus destroyed all our documents."[20] That's what corporate top brass expect when they tell their underlings to destroy incriminating documents. But the news that Enron Corporation had shredded documents after it received a notice that the company was under investigation came as a shock not so much about the shredding itself but more so because it had been taken for granted that big corporations had moved to electronic accounting and record keeping.

One could see the dubious wisdom of keeping records in the old-fashioned way; that is, as paper documents, so that a corporation could destroy them, and then, perhaps, hire some smart attorney to cook a plausible explanation for public consumption. One cannot retrieve a shredded or a burned document, but if an electronic document is deleted, it can be recovered even if the computer is not networked. In future, corporations would use special software to make their dubious electronic documents disappear in case the FBI or Internal Revenue Service became too nosy. To do so, of course, they would have to transfer their information to some safe haven, some off-shore anonymous data Cloud warehouses, much the same way as some super-rich people put their money into Swiss bank accounts, Cayman Islands' shell companies, or some place where neither the free press nor the American law operates to make them accountable. Perhaps such a place has yet to be invented in the aspirational society.

HIT THE NUMBERS OR YOU ARE OUT

The unceremonious firing of Carly Fiorina, once upon a time the glamorous, globetrotting CEO of Hewlett Packard, one of the few women who broke the proverbial glass ceiling—some would say by sheer guts and grit—was an odd occurrence in the annals of corporate America.[21] Boards of directors normally don't act that brazenly; they sweeten the departure with friendly negotiations, mutual understandings, and hefty severance packets, which sometimes seem so excessive that they evoke public repugnance. Fiorina, too, got a hefty purse, a golden parachute of $21 million, but she was fired because of her inability to beat the competition rather than any fraud she might have committed.[22] In the United States, fraud and incompetence are two sides of the same coin; they hurt the market as well as the shareholders.

The functioning of corporate America is based on authoritarianism, not on internal checks and balances. In contrast, the U.S. political system is based on a healthy distrust of people in power. The tripartite system of government of coequals—the president, Congress, and the Supreme Court—and other built-in checks and balances along with a free press have kept politicians from abusing power by exposing them to the threat of public ridicule, impeachment, or jail.

Think of Richard Nixon and hosts of governors and legislators who have been disgraced. The founding fathers of American democracy didn't leave the functioning of the political system to the innate goodness of the people in power. Nor did they leave the development of good political behavior to any kind of special training in schools or colleges, or the culture of the sports arena, for that matter.

The temptation of power trumps everything else, so transparency and accountability are indispensable for the aspirational society. That unfortunately isn't the case with Wall Street, where most Americans are invested through pension and other retirement accounts. George W. Bush's idea of an ownership society would have invested future social security funds into the stock market with unforeseen consequences.[23] Today we live in a world where corporate power overshadows most of our activities, and therefore it needs constant vigilance and strict scrutiny by the press, which must remain independent, bold, and fearless. When journalists remain neutral or are incapable of understanding the workings of Wall Street or collaborate with its functionaries for personal favors, Main Street suffers.

THE MORAL NEUTRALITY OF NUMBERS

Corporate leaders rise to power on the promise of maximizing profits, market value, and economic growth for their companies. There are no internal checks and balances in the form of a healthy opposition seeking accountability and presenting an alternative vision of the company.

Shareholders are passive, selfish, and shortsighted. Boards of directors are morally neutral; their interest is limited to dividends, capital gains, perks, and other forms of payoffs. They worship executives who maximize their investments. As long as a CEO performs well and exceeds the expectations of Wall Street, they could get away with some excesses. Fiorina was fired for a style of management that despite its aggressiveness did not produce results to cheer the market. Fiorina lacked imagination—the power to do "super-speculation" about an alternative future for Hewlett Packard when under her leadership the merger with Compaq did not work out as expected.

But other CEOs have done worse; for example, former chief executives Bernard Ebbers of WorldCom, whose $11 billion fraud dumped the telecommunications company into bankruptcy. The former financial chief of the company, Scott Sullivan, who cooperated with prosecutors after pleading guilty and testified against his former boss, told the jury that he had warned Ebbers that accounting adjustments, creative accounting or cooking books, could not be justified. Ebbers told him nevertheless that the company had to "hit the numbers" and meet the financial and revenue targets. Ebbers blamed his underlings for the fraud, said to be one of the largest in U.S. history, for which he was convicted and sentenced to prison for twenty-five years. At its peak in 1999, WorldCom had a market capitalization of $180 billion, and Ebbers was the toast of Wall Street. That was the problem. To maintain its market reputation WorldCom had to cook books and lie. When WorldCom's real earnings could not meet the forecast, Ebbers asked the account department to "adjust the numbers."

Accounting departments are notorious for doping and duping until they are caught. Wall Street analysts and financial journalists who, out of fear or favor, work as paid employees of big corporations rather than watchdogs of public interest went along with the web of lies woven by the WorldCom team until the whole edifice began to collapse in 2000 and the share price tanked to $15 from a high of $64 in 1999.[24] Ebbers's personal fortune, too, was linked with WorldCom's market share price, and in order to keep it high, he raised the analysts' expectations.

Wall Street is an unforgiving and cruel taskmaster. One cannot get away with lies for too long in the aspirational society predicated upon openness. But sometimes the price a company, and eventually the public, pays is too high, and the damage is irreparable.[25] But can we make corporate bosses honest? Can we instill ethics into their consciousness?

Robert J. Shiller, a Yale professor and author of *Irrational Exuberance*, faults the education that U.S. business schools impart to budding executives. Writing in the *New York Times*, he said that instead of emphasizing ethics and liberal arts, "modern business education often encourages excessive respect for anything that can be considered a result of the free market."[26] Teach them ethics, he thundered quietly.

The average age of a student in a U.S. business school is twenty-eight, and by that time shouldn't a person be able to know what is not ethical? Don't depend upon liberal arts and ethics education alone, because that is not enough. Transparency and internal checks and balances, including a whistle-blowing cell, must be embedded into the system. More importantly, if the news media pay as much attention to the internal workings of corporate America as they do to the scandals of political America and sexual escapades of Hollywood, corporate CEOs would behave better and the global financial future would be secure.

MONEY, MONEY, MONEY, AND MORE MONEY

Corporations exist to make money, not to spread the ideology of free marketplace capitalism, freedom of speech, or democracy. One of the best ways of making money is to dominate the market by eliminating competition. So corporations expand through mergers and acquisitions, or they make unique products that give them exclusive positions in the marketplace. When they grow big and rich, it becomes difficult to keep their capital bottled up in one place. Some of it flows into the pockets of the people who control them. These people become victims of their own success and meet their hubris.

Arthur Levitt Jr., former chairman of the Securities and Exchange Commission, lamented in the *Wall Street Journal* that "exorbitant compensation feeds the worst instincts and egos of powerful CEOs, fuelled by their desire to win at all costs and resulting, too often, in the cutting of ethical corners."[27] Levitt blamed the prevailing corporate boardroom culture of interlocking relationships, of "you scratch my back, and I yours," for unseemly greed. And he wondered, "How likely is that a board member would challenge the person who invited him to join, and can re-invite him? How likely is it that a board member would stand up to the CEO who directed the company's foundation to support his favorite charity? Or hired his law firm? Or his wife's interior decorating shop?"[28]

Greater corporate board independence and accountability and public exposure by vigilant media about the rampant, excessive, and outrageous executive compensation might offer some hope. In the era of resurgence of marketplace values, how much does a company executive deserve? Or what punishment would he deserve if he pulled the company down the drain? But in the age of globalization, shouldn't our sights extend to the roaring economic giant China? If a Chinese business goes bust, can the United States turn a blind eye? Consider the case of Chen Jiulin, the Chinese entrepreneur who used his country's mystique of flying dragons and crouching tigers,

"Chinese Wisdom and International Expertise," to turn China Aviation Oil (Singapore) Corporation into a most sought-after investors' sweetheart. Millions of dollars of investors' money poured into the company, and the stock rose by 80 percent. But then the truth was out. On November 25, 2004, Chinese Aviation Oil bent its knees before a Singapore court and sought protection from its creditors.

The miracle company had lost $550 million in speculative bets on oil. But the state-owned parent company in Beijing knew the unfolding disaster and tried to stem it by selling 15 percent of China Aviation Oil (Singapore) to investors to raise money to cover the losses, however, without full disclosures to the investors.

In the United States, this kind of hush-hush sale is called insider trading, the kind of illegal activity for which many American CEOs, including the media darling Martha Stewart, served jail terms.[29] Of course, Chen Jiulin, the East-West business philosopher, the suspended CEO of China Aviation Oil (Singapore), who ran away to his village in China where he grew up as a farm boy, was brought back to Singapore to facilitate the inquiry into the biggest collapse since the Barings Bank in 1995, which suffered a loss of $1.2 billion. "The scandal has raised questions about corporate governance in China," wrote Clifford Coonan in the *Times* (London), which should have been a matter of serious global concern and more so when the world is rushing to invest in Chinese companies without knowing how hollow are those outfits.[30]

By controlling information and hyping their prospects, Chinese companies blow balloons that can't be sustained. So when a far-off company, beyond the control of American regulators, goes bust, who do you call? China Aviation Oil (Singapore) was simply an arm of the parent Chinese-government-controlled company, which had created a monopoly over the aviation oil supply to China. Even outside its borders, China tries to create conditions of complete command-and-control by creating cartels and monopolies. A Singapore court convicted China Aviation Oil's Jiulin, but that was mere shadow fighting. What can the Singapore or any other government do about the Chinese government that spreads its tentacles abroad through privately held shadow companies? Absence of freedom of the press in China is dangerous for the world, a topic we shall explore in the next chapter.

WHEN DRUG COMPANIES SELL SNAKE OIL

It will be hardly an exaggeration to say that global drug companies might have killed more people than gun violence, road accidents, or even terrorists. In his testimony before the Senate Committee in 2004, Raymond V. Gilmartin, former president and CEO of the pharmaceutical giant Merck & Co., said that even his wife took Vioxx until the day it was withdrawn, the arthritis painkiller that for some people became a killer drug.[31] Cynics might say that Gilmartin did not care as much for the American patient as he did for the shareholders of his company.

In an open letter Gilmartin said that the company "is continuing to offer to refund patients for their unused Vioxx."[32] But Gilmartin was silent about what kind of refund Merck would give to children whose mothers have been killed by Vioxx. After a study established that patients taking the arthritis drug Vioxx for eighteen months had twice the risk of heart attacks and strokes than those placed on a placebo, Merck immediately pulled the drug off the market. However, memos and emails leaked from the company to the media indicated that Merck had some prior knowledge about the side effects of the drug long before it was taken off the shelf. Did the company use the glitter of its corporate power to choke the bad news?

The U.S. Food and Drug Administration, which approved the drug in 1999, reported that Vioxx might have been responsible for causing an additional 27,785 heart attacks or deaths from 1999 through 2003. Shouldn't the FDA be held culpable for not supervising the drug properly? Dr. David J. Graham, the researcher who led the Vioxx study for the FDA, told the Senate Finance Committee that Vioxx was a "national disaster" and that the FDA embodied "a profound regulatory failure" and was "incapable of protecting America against another Vioxx."[33] The drug, Dr. Graham said, should have been removed from the market four years ago, alleging that some people in the company knew that the drug was not very safe but kept quiet. This was nothing but reckless disregard of the truth, what one might call actual malice against the public.

What Ralph Nader said about the auto industry in his book, *Unsafe at Any Speed*, is more or less equally true of the pharmaceutical industry: Can any drug be totally safe?[34] What are the FDA's safety standards? Sandra L. Kweder, the FDA's deputy director in the Office of New Drugs, said, "Unless a new drug's demonstrated benefits outweighs its known risks for an intended population, FDA will not approve the drug. However, we cannot anticipate all possible effects of a drug during clinical trials that precede approval."[35]

But Dr. Graham in his testimony before the Senate Finance Committee compared FDA drug warnings with weather forecasting.[36] He told the Committee that if a weather forecaster said that there's an 80 percent chance of rain, naturally a person would take an umbrella to his office. But the FDA believed that an umbrella wouldn't be necessary unless there is a 95 percent probability of rain. The analogical criticism of the FDA, however interesting, did not answer the question of acceptable risk for a drug. If a drug kills five patients but saves or makes the lives of a million livable, should that drug be put on the market? Should the FDA obligate the pharmaceutical company that manufactures the drug to put a label on the drug, such as "five in a million users of the drug will die or develop some other serious problems?"

Many people wonder whether the FDA has become a promotional arm of the pharmaceutical industry or is intrinsically incapable of doing its work properly. Dr. Jerry Avorn, associate professor of medicine at Harvard Medical School and the author of *Powerful Medicines: The Benefits, Risks and Costs of Prescription Drugs*, talked with Ray Suarez of *PBS NewsHour* about whether the FDA was doing a good job of protecting the public. Dr. Avorn said,

> I think it's fair to say in light of both the Vioxx experience and also the experience we've had with other drugs in the recent past that, no, the system is not working well at all. There's an enormous focus within the FDA about approving drugs quickly and getting them on to the market. And that's okay if it's done well. But then the attention of the FDA really drops off. And the vigilance disappears when it comes to requiring and analyzing the data that we need to be able to learn about the safety of the drugs once they're in widespread use.[37]

It is difficult to say whether the FDA suffers from a short attention span or is intimidated by corporate power, but Congress and the media can't ignore the problem of drug safety any longer. Unfazed by criticism, Merck conducted a massive public campaign to win the hearts and minds of its shareholders if not the American public. The three full-page ads in major newspapers in 2004 were not meant to assure the millions of people who used Vioxx. Like the rest of corporate America, Merck is interested in the bottom line, and that's what Gilmartin said: "Our business prospects are strong and we are well prepared to address the challenges posed by the withdrawal of Vioxx. We have a strong balance sheet, with cash and reserves that well exceed debt. We have high performing products in major franchises that we expect to generate the cash flows we need to fund research, build new manufacturing capacity and continue to deliver returns to shareholders."[38]

There is nothing obscene about Merck's interest in delivering "returns to shareholders," provided it also includes the welfare of the American patient. Merck stood on trial in the court of public opinion and before Congress and faced class action suits, and after seven years of litigation the company finally reached a settlement in 2011, pleading guilty to a criminal misdemeanor charge and agreeing to pay $950 million as penalty.[39] Now it would be Merck's obligation to prove what the ad claims: "Our ethical standards are the foundation of our company. Merck has consistently been recognized as one of the world's ethical companies."[40]

One might wonder as to who bestowed such a high distinction upon Merck? Perhaps they are confusing ethics with technical efficiency and profit making. Raymond V. Gilmartin, now an adjunct professor at Harvard Business School, should be having enough time to reflect on his own controversial executive performance at Merck and be able to dispense fresh advice and ethical insights to his students. Writing in his blog, he said, "In the past 25 years, CEOs of many major corporations have relied on a flawed set of beliefs to lead their organizations. This set has influenced them to place way too much emphasis on maximizing shareholder value and not enough on generating value for society."[41]

FREE PRESS AND CORPORATE BEHAVIOR

Free marketplace capitalism is an amoral, albeit an extremely powerful, force for creating wealth and prosperity. But left unsupervised, however, it can cause global havoc, as happened during the 2007 to 2008 financial meltdown. Lehman Brothers collapsed, Bear Stearns was folded into JPMorgan Chase, and the federal government had to bailout several banks and corporations thought to be too big and indispensable, whose downfall would have pulled down the whole economy into a deep depression, the worst since the Great Depression. Corporations have short-term memories and no social conscience. Few corporations learn from their past mistakes.

Left alone in their self-created web of interlocking ownership and operations and financial and accounting complexities, corporations and their executives will be tempted to cheat and commit fraud, as the two most recent cases, GlaxoSmithKline and Barclays, demonstrate. Their criminal behavior was not unprecedented or unique. It happened before and will happen again regardless of how rigorous the regulations, how ferocious the watchdogs.

GlaxoSmithKline pleaded guilty to federal crimes and settled to pay approximately $1 billion in fines because it had promoted, in collusion with physicians, two prescription depression medications Paxil and Wellbutrin for unapproved or off-label uses; and a third one, Avandia, a diabetes drug about which it had failed to report clinical safety data to the FDA. Moreover, GlaxoSmithKline, a UK pharmaceutical company, agreed to pay an additional $2 billion to settle civil claims filed by both federal and state authorities alleging that the company paid inducements to physicians to prescribe drugs, apart from making false and deceptive statements about drug safety and defrauding the Medicaid drug rebate program.

GlaxoSmithKline should have learned a lesson from Pfizer, Eli Lilly, Johnson & Johnson, and Merck, who had pleaded guilty to similar charges and entered into settlement with the Justice Department. To GlaxoSmithKline's dubious distinction, its $3 billion settlement was the largest health care fraud settlement in history.

Barclays, a three-hundred-year-old British bank, was charged with the manipulation of the London interbank offered rate, or LIBOR, a benchmark rate indicator that determines the global flow of capital and forms the basis for trillions of dollars of financial products, including mortgages, savings, student loans, and complex derivatives. In June 2012, Barclays agreed to pay $450 million to British and American authorities to settle claims that the bank had rigged the number, though Barclays was not alone in the rate-rigging game.

A cartel of eighteen banks' reports to the British Association of Bankers in London estimated rates they would be charged to borrow from other banks for a variety of short-term maturities in currencies, including dollars, euros, and yen. LIBOR is the average of the reported rates, but it is based on estimated rather on actual transactions. The lower the LIBOR, the stronger is it supposed to be the banks' financial health and greater the trust. During the financial crisis of 2007 to 2008, Barclays and other banks rigged their borrowing rates for the benchmark LIBOR, which according to the *Economist* "turns out to have been flawed."[42]

Most importantly, the behavior of GlaxoSmithKline and Barclays has been no different from other companies such as Enron, Global Crossing, Tyco, WorldCom, and Merck, as discussed earlier in these pages, and that of JPMorgan Chase that reported trading losses amounting to billions of dollars, which were on flawed bets, similar to the kind of complex derivatives that played such a devastating role in the financial crisis.[43] As mentioned earlier, corporations have short-term memories that make them feel invincible.

Ben W. Heineman Jr., a former GE executive and a now a senior fellow at Harvard, argued in *The Atlantic* that the Barclays and GlaxoSmithKline scandals show that a hands-off approach to markets does not work, and hence the challenge of "how to realize the benefits of market capitalism while restraining the powerful impulses to cut corners, cheat, and commit fraud."[44]

Summing up the corporate conundrum, Heineman wrote, "Corporations large and small apply relentless internal pressure on their people to hit basic financial goals for net income, cash flow, and stock price. Other commercial targets, too, may be critically important: achieving specific returns on investment, equity. or assets; hitting sales or service goals; meeting product development or product launch schedules; and attaining productivity increases. Companies, especially those in the global economy, also face external pressures such as endemic corruption, weak rule of law, pervasive conflicts of interest, and extortionate demands from government officials."[45]

Heineman suggests a two-pronged, integrated approach to regulate corporate behavior. Corporations should have an internal integrity program that establishes a culture of creativity, innovation, and high performance based on high ethical and legal standards. But since such a culture of integrity and ethical behavior cannot exist by itself alone, it has to be sustained by outside agencies, "broader societal rules," because "without the threats and sanctions of public law in many areas of the economy, most companies would not seek to create a culture that seeks to constrain wrong-doing by their employees."[46]

But such regulatory approaches already exist in every corporation. For example, Barclays, like any other financial institution, has been regulated, and its website blithely crows: "Barclays has a clear sense of its business purpose—to help individuals, communities, businesses and economies progress and grow."[47] Like other pharmaceutical companies, GlaxoSmithKline operates under strict rules and regulations, and its proclaimed corporate social responsibility is laudable. As CEO Andrew Witty states, "At GSK, we firmly believe that operating in a responsible and ethical way is essential for the success of our business. We have come a long way but recognise that there is much more we can do and we will continue to challenge ourselves."[48] In spite of rules and regulations, these are the two most recent global corporations that have committed high crimes and misdemeanors. Will more rules and regulations make corporations better citizens now that the Supreme Court has recognized their First Amendment free speech rights under the *Citizens United* landmark ruling? Something more is needed to regulate the behavior of corporations.

WHO DO YOU CALL WHEN NOTHING WORKS?

In the wake of Enron-WorldCom scandals and the economic and financial crisis of 2008, several scholars and corporate executives have been pondering over the question of corporate governance and how to restore faith in free marketplace capitalism. Now more than ever, business schools emphasize the teaching of ethics, corporate social responsibility, and the obligations of corporate citizenship. Corporate governance is also the focus of many think tanks.

Consider the Committee on Economic Development (CED), an independent, nonpartisan American think tank, which is peopled by senior corporate executives and academic leaders. According to its mission statement, CED is "dedicated to policy research on the major economic and social issues of our time and the implementation of its recommendations by the public and private sectors." In its January 2010 Policy Report, authored by Ben Heineman, "Resorting Trust in Corporate Governance," CED recommends six cohesive steps that board of directors and senior executives must take to build long-term, value-based, strong economic performance, well-thought-out risk management, and unimpeachable honesty.

The CED Report asks corporate America: (1) to redefine its mission, the role of the board of directors and CEO in order to establish a new balance between risky ventures for innovation and growth and the sound management of such risky behaviors so that emphasis should shift away from shareholders' value to long-term growth and societal values; (2) "performance, risk, and integrity" should be the guiding values for the training of corporate leadership; (3) for this redefined corporate mission, the CEO selection process should focus on appropriate leadership skills; (4) a new set of financial and nonfinancial metrics should be developed for the assessment of corporate leaders on "performance, risk, and integrity" in relation to long-term growth that's inclusive and how it impacts society; (5) a significant part of CEO compensation, cash and equity, should be variable and be based on how far the objectives of "performance, risk, and integrity" have been met; (6) the board of directors' top priority should be how far the corporate mission of high "performance, risk, and integrity" has been accomplished.[49]

The CED "Six Essential Tasks" corporate governance report received wide publicity in the media and academic circles, but like myriad similar reports that sprout after every major crisis, it has had as little effect upon corporate governance as *Sesame Street*'s "Monsterpiece Theater: 39 Stairs" has had on children's ability to learn numbers—if one considers the two most major scandals, GlaxoSmithKline's unauthorized, off-label drug peddling and the Barclays cartel LIBOR rate manipulation.[50]

In spite of heavy fines, impaired reputations, and hefty jail sentences, corruption in business continues and bribery persists, according to Ernst & Young's 12th Global Fraud Survey.[51]

Democratic free marketplace capitalism based on open competition is indispensable for continuous economic growth, innovation, creativity, and spreading prosperity to the base of the pyramid in the aspirational society. Corruption is the enemy of the free marketplace because it distorts markets and creates unfair competition. Often companies, from energy to health care, are incentivized to pay bribes and rig bids to win public works procurement contracts.

Tactics used by companies to abuse open competition include secret subsidiaries and partnerships, exploiting legal loopholes and tax laws, and forming hidden cartels. According to a Gallup poll, 62 percent of Americans believe that corruption is widespread in businesses.[52] Ironically, in spite of the widespread perception of corruption, the World Bank Ease of Doing Business Index for the United States is 4. Doing business is, however, not the same as people having trust in Wall Street, banks and financial institutions, and pharmaceutical, oil, and other multinational global companies.

Why does corporate corruption persist in the United States and other developed countries in spite of strong political leadership, well-thought-out policies, laws, and greater transparency? The obvious answer is that corporations believe that they are indispensable, that they are too big not to be salvaged by the government if they get into trouble, that they can get away with corruption by hiring the best legal brains and lobbyists that money can buy. What is a $3 billion settlement for a global pharmaceutical giant like

GlaxoSmithKline? Barclays bank chief executive Bob Diamond gave up bonuses worth $31 million after he resigned in the wake of a rate-rigging scandal, but Barclays' practices will continue.[53]

Who do you call when nothing else works? Since holding corporate CEOs responsible for corporate corruption doesn't work, the board of directors severely and collectively should be held liable for corporate corruption. To protect themselves from public calumny, corporate boards should establish the equivalent of the Homeland Security apparatus to preempt corruption at the aspirational level rather than discover corruption when it has already happened.

Since the traditional news media organizations are either incapable of understanding the workings of Wall Street firms, the complexities of financial instruments, the intertwined operations of global corporations, or are prone to collusion with them through bribery, insider trading, careerism, or intimidation, the public has no choice but to welcome and embrace a free-floating, footloose, beyond-national cyberorganization such as WikiLeaks.

Chapter Six

China's Authoritarianism Challenges American Democratic Exceptionalism

THE ROAR OF THE DRAGON

Remember the time when European and American corporate diplomats used to crow, "You cannot afford not to be in China."[1] Keeping up with the newly emerging field of global corporate diplomacy, American CEOs, instead of admonishing China for its egregious behavior, have been overly eager to apologize for their customers' complaints.

In 2007 the world's largest toy maker, Mattel, for example, recalled twenty million lead-paint-tainted toys manufactured in China, but instead of reprimanding its Chinese vendors and looking for an alternative contract manufacturing source, its executive vice president, Thomas A. Debrowski, delivered a demeaning, pathetic apology: "Mattel takes full responsibility for these recalls and apologizes to you, the Chinese people, and all our customers who received the toys."[2] It is just like a rape victim apologizing to the rapist, "Pardon me for tempting you, Sir," or as Senator Charles E. Schumer said, "It's like a bank robber apologizing to his accomplice instead of to the person who was robbed. They're playing politics in China rather than doing what matters."[3]

Cleverly staged mass media propaganda and lobbying by people in high places, including some of the top CEOs of major U.S. corporations, former diplomats, and academics have helped the Chinese authorities in blurring facts with fiction, creating the perception of China's relentless and inevitable rise to a global superpower. China has been fascinating corporate America with the state-created myth of bountifulness of its high-skilled, disciplined, cheap labor, but more so with the Chinese Communist Party's ability to control the obedient masses that produce low-priced goods for corporate

America's global markets. Besides, China's proliferating middle class would one day buy every branded product made in the United States. China, too, has come to believe the constant drumming that Americans cannot do without its inexpensive goods. So why worry about intellectual property and trade secret thefts, currency manipulation to boost its exports, humongous trade surpluses, fakes and counterfeits, defective and tainted toys, and contaminated food products?

The 2008 Beijing Olympics was touted by the International Olympic Committee to be an event that would reshape the sports and marketing spectacular extravaganza into a force that would trigger political and social changes in China. That was expecting too much from an organization like IOC that has been paying little attention to its own widespread problems, including bribery scandals and drugs.[4] If human rights were the deciding factor in determining the choice of the host city for the Games, Moscow under the Soviet Union and Berlin under the Nazis would not have been selected to host the Olympics.

China was selected to showcase the Games despite its abominable record of the suppression of human rights of the people of Tibet, followers of the banned spiritual movement Falun Gong, political dissidents, and scholars rotting in its jails without recourse to a fair trial.[5] Slavoj Zizek, a Slovenian philosopher and cultural critic, wrote that in Tibet, the Chinese authorities "in addition to military coercion, they increasingly rely on ethnic and economic colonization. Lhasa is transforming into a Chinese version of the capitalist Wild West, with karaoke bars and Disney-like Buddhist theme parks . . . in a decade or two, Tibetans will be reduced to the status of the Native Americans in the United States."[6] He should have included gambling casinos, drugs, booze, and obesity of the Native Americans, a possible dreadful future for Tibetans also.

VICIOUS CAPITALIST ENERGY

Doing business with China is more important than human rights, though Americans, along with rest of the world, go on paying lip service to the problem.[7] Trade and the Olympics had no civilizing effect upon Nazi Germany or the Soviet Union; therefore, to expect a miracle to have happened in China because of the Olympics in 2008 was puerile expectations. Rising prosperity will not force China's Communist Party to give up its monopoly over power and become democratic. Since Deng Xiaoping took the road to state-controlled capitalism, replacing communist brutality with capitalist ruthlessness, China's economy has been opening up and growing rapidly, with its gross national product (GNP) rising to more than $7.30 trillion in 2012, making it the second largest economy today. For the past three decades, the rate of annual economic growth has hovered around 9 to 10 percent. Made-in-China goods, apple juice, toys, shoes, electronics, golf clubs, and fake luxury brands are found in every shopping mall of the world. Huge economic benefits were expected from the 2008 Games because it necessitated the investment of billions in infrastructure and information technology to modernize and showcase Beijing for the events. Millions of tourists poured into China, and some took a train to Lhasa and visited the Three Gorges superdam; nonetheless, they remained silent observers.

The Deng Xiaoping market economy revolution unleashed China's vicious capitalist energies, but not without the help from the outside world, especially the United States, which magnanimously opened its markets to China. Today China is a healthier, better-fed, and better-educated nation than most other developing countries, but it remains a closed society ensconced in pollution and fear of domestic instability. Who needs freedom, if China can make so much progress without it?[8] That's why George W. Bush never preached freedom to China, though he repeatedly asked that it should let its currency rise. Nor does Barack Obama—whose kowtowing attitude at the 2010 nuclear security summit meeting with president Hu Jintao left everyone puzzled—care for freedom and human rights in China.

Today China feels confident that it can compete and will soon surpass the United States. But it does not have to tolerate the noise and chaos, the perpetual turbulence, of an open, multicultural, and multiracial messy society such as the United States, where the people demand accountability from their political leaders. Beijing, with the help of somewhat reluctant Google, Yahoo!, Microsoft, and Cisco, has been expanding its control into the digital domain and eventually aims to have the same control over cyberspace as it has over Tibet and Xinjiang. Chinese authorities know that left to itself or to the Americans, the Internet might bring about tremendous political upheavals in China, as it has been doing in Middle Eastern countries where long-standing dictators have been blown away like leaves.

Large, centralized political systems have a tendency to break down due to internal pressures triggered by communications technology, unless they have built-in capabilities for adjustment, which China does not have at present. It is difficult to say what might happen in China in the age of the Internet, satellites, cell phones, YouTube, and hosts of other wireless, digital, and interconnected sensing devices that are becoming available to everyone.

China wants to control the uncontrollable, the digital generation swapping billions of text messages on cell phones, the generation that could self-organize itself into an uncontrollable smart mob. If the 2007 Myanmar Buddhist monks' uprising did not disturb the rising superpower, the Middle East uprisings, euphemistically called the Arab Spring, that began in January 2011 in Tunisia and is continuing in Egypt today and has turned into civil war in Syria, must be giving the Chinese rulers sleepless nights.

The authoritarian regime temporarily opened its doors, skies, and cyberspace to a worldwide audience during the 2008 Olympics, but once the crowd was gone, the sky dome closed, of course, with the help from corporate America, which one day might become a dancing bear for China, if others follow Mattel's obsequious behavior.[9]

EXTREME NATIONALISM

China seems to be a nation that is intoxicated with great expectations about its future, while the rest of the world is just besieged with bad debts, spiraling prices, and random eruptions of domestic violence and terrorism. From die-hard communism to state-controlled marketplace capitalism, it has been a long march indeed, though not what Mao Zedong might have conceived. But that is in fact a great tribute to China's genius: its adaptability and resilience, creating the perception of China's relentless and inevitable rise to a global superpower without robust political freedom as we know in the United States. A 2012 Pew Global Attitudes Project survey showed that an overwhelming majority of Chinese said they were happy with their country's direction provided by the Communist Party, in spite of growing concern about inequality and corruption.[10]

China fascinates the global corporate with its controlled narratives of boundless opportunities and more so with the power of its ruling party's collective will that is capable of directing millions of hardworking, entrepreneurial, and yet obedient masses. Doing business with China is more important than human rights, though Americans, along with the rest of the world, go on paying lip service to the issue. In 2008 the U.S. House of Representatives passed a near unanimous resolution (419 to 1) criticizing China's human rights record, but since then Congress has been muted in its criticism.

Since China took the road to capitalism about three decades ago, its historically unprecedented economic growth has made the Chinese, especially its elite and entrepreneurial classes, ultranationalistic and patriotic. Tiananmen has been wiped out from the nation's historical memory. The rest of the world, too, has begun to see new opportunities in China.

Some U.S. corporations cannot stop thinking that by offering a selective partnership to Chinese businesses they would be able to share, if not totally co-opt, China's brain power. For example, after selling its ThinkPad to a Chinese company, Lenovo Group Ltd. in 2005, IBM alerted the public about the inevitability of China's rise and the need to harness its strength for corporate America. A full-page advertisement amusingly admonished, "The future is a dragon. Do you hear it coming?" IBM boasted about its access to a global pool of Nobel laureates, research labs, and no less than three thousand scientists, engineers, and technologists. Instead of paying the salaries of scientists and technologists to solve complex problems, the ad asked, wouldn't it be great simply to rent their brainpower?

Harnessing China's brainpower for doing specific jobs may sound more acceptable than the outsourcing of jobs, but post-Olympics China's intellectual and manufacturing power may no longer be available for renting. The Japanese, too, have been hearing the rise of the dragon. In 2005, the Chinese government permitted and encouraged loud and sometimes violent protests against Japan in several big industrial cities, including Shanghai and Hong Kong, regarding the Japanese insensitivities to Chinese bruised feelings, as it happened again over disputed islands in 2012.[11] The Chinese claimed that their feelings had been hurt because some Japanese school textbooks showed no regrets about the atrocities the Japanese troops had committed against them during World War II. There were other reasons. Japan had begun to explore undersea oil and gas deposits in a disputed region of the East China Sea. And of course Japan's continuous strategic alliance with the United States regarding the Taiwan issue has been always an irritant. When Japan asked for an apology and compensation for vandalism and damage to its diplomatic and commercial property, China said it had nothing to apologize about. Before the street protests, the Chinese government had allowed an online petition drive by millions of Chinese against Japan's effort to seek permanent membership in the UN Security Council.[12]

The unprecedented online phenomenon showed how China could mobilize its masses. Just as the Chinese authorities aroused the Chinese to come out and protest against Japan, with the same speed they ordered protesters to disappear. At present it seems that the Communist Party is capable of generating and controlling mass enthusiasm through nationalism. But the Chinese-controlled nationalism does not answer the question: If China cannot export what it manufactures, what will it do? Furthermore, if China has no place for investment or depository for its billions of annual trade surpluses, what will it do? But these are questions for the future. Why worry now when China has been growing at an astounding rate and has become the second-largest economy in the world and, moreover, expects to become a great military power in the next decade or so? Since the one-party authoritarian rule has not hampered China from growing at an extraordinary rate, it is reasonable to ask if China could accomplish so much in such a short time without civil liberties including political freedom, transparency, and freedom of the press, couldn't other countries do the same if they follow the Chinese model of one-party, elite rule that governs like a corporate board?

NO GRATITUDE

China is one of the United States' biggest foreign buyers of its Treasury notes, but it earns no respect from Americans; instead, it creates fear. Gao Xiqing, president of the China Investment Corporation that managed about a $410 billion sovereign fund, told James Fallows of *The Atlantic*, "Be nice to the countries that lend you money."[13] In other words, don't talk about human rights, Tibet, piracy, currency manipulation, bad toys, milk products that kill children, or cyberespionage.[14] Many American politicians, including former Speaker Nancy Pelosi, have been great critics of China's violation of human rights and unfair trade practices. But their rhetoric of idealism is compromised by the international realities of China's rising economic power.

When China says it might diversify its foreign exchange holdings, the dollar shudders a trifle. China holds more $3 trillion in foreign currency reserves (a substantial portion in U.S. Treasury notes) thanks to its export-driven economy and controlled currency value. Japan, too, has massive foreign currency reserves, more than $1 trillion, but the United States has never felt threatened by it. Moreover, human rights and freedom have seldom been an issue between the United States and Japan or pro-American Muslim-Arab countries.

As a global power, the United States cannot give up dealing with authoritarian regimes regardless of its sporadic zeal to spread freedom universally. But China is different because it is an economic superpower, and its global trade without freedom of information and the press is likely to have adverse effects on rest of the world. The talk of the absence of freedom of the press and civil liberties in China is not a tool of public diplomacy against China. The most important question is whether there can be free and fair global trade in the absence of a free and fearless press necessary to watch China's global transactions, including the safety of products made in China.

Most Americans believe that the United States would remain vulnerable to terrorism so long as authoritarianism and hate ideology prevail abroad, and to beat terrorism there is no other solution except to expand freedom and openness. But that is equally true of China, albeit China is doing well without democratic freedoms. Day and night China's disciplined and hardworking people make goods for the entire world without much thought about freedom, and many countries in the developing world look to China not only for aid and trade but also as a wonderful working model. The biggest challenge for U.S. diplomacy today is to persuade China to open up and loosen its authoritarian control over society because transparency is essential to building global brands, if China wants to move beyond contract manufacturing.

China's peaceful transformation from communism to state-controlled, export-driven capitalism is in fact a great tribute to China's pragmatism, creating the perception of China's relentless and inevitable rise to a global superpower. China has been attracting massive foreign direct investment (FDI) with its managed narratives of boundless opportunities, and more so with the power of its ruling party's ruthless collective will that it can accomplish anything. China believes that since the world cannot do without its inexpensive goods and talents, there is not much to worry about intellectual property, currency manipulation to boost exports, massive trade surpluses, and rising foreign exchange reserves that end up as U.S. Treasury notes. Rising prosperity has not encouraged China's rulers to loosen their control over power. China feels that it can compete with the best without facing problems that an open society like the United States confronts. But China might have reached a stage of development when freedom becomes an economic necessity.

Some U.S. corporations naively think that by offering selective partnerships to Chinese businesses they would be able to tap into China's brainpower. On the other hand, the Obama administration has concluded that borrowing money and brainpower from China to sustain U.S. global power is not an option any longer. A few years ago, China stunned the world when it destroyed one of its own aging weather satellites by hitting it with a ballistic missile five hundred miles into space, thus signaling its intentions to master space. The United States cannot ignore such intentionality, even though China has ceased to be an imminent threat since its economic growth has become increasingly tied up with the search for energy and other raw materials, foreign direct investment, and exports.

Through its growing trade surplus and currency reserves it seemed until recently that China had established an iron grip on the United States. But the 2008 financial crisis and the recession, contrary to the general impression, have energized the United States. President Barack Obama is determined to rebuild the $15 trillion economy through innovative strategies for economic self-reliance (not protectionism). As China scrambles to reshape its export-based economy, it will face a new United States of America—bolder, smarter, and more imaginative. The beat of the American drum will be heard around the globe for decades to come. There is no end to the aspirations of the aspirational society predicated upon the culture of challenge and response, free speech, and innovation.

PACIFIC OCEAN ISSUE

Mark Helprin, a senior fellow of the Claremont Institute for the Study of Statesmanship and Political Philosophy, contends that China is following the example of the militaristic Japan under the Meiji era, and it plans to dominate the Pacific Ocean. China "harbors major ambitions" and plans to counter the United States in outer space, oceans, and in cyberspace. While the United States is bogged down in fighting terrorism and domestic gun violence, China is slowly taking the weaker states of Southeast Asia (consider, for example, the ASEAN-China Free Trade Agreement) under its protective wings. Helprin warns, "This century will be not just the century of terrorism: terrorism will fade. It will be a naval century, with the Pacific its center, and challenges in the remotest places of the world offered not by dervishes and crazy-men but by a great power that is at last and at least America's equal. Unfortunately, it is in our nature neither to foresee nor prepare for what lies beyond the rim."[15]

When the *Shenzhou 5*, "Divine Vessel," spacecraft landed safely in the grassland of Inner Mongolia, China flushed with pride, and justifiably so because managing a complicated project like the launching of a manned space flight requires not only the physical training of astronauts but also the meticulous technology capabilities of mastering man-machine functioning and coordination of so many factors, some not under total human control. China is only the third nation to fly human beings into space and bring them back safely to earth. China has taken a great leap forward. It is much more than reinventing the wheel. Such technological achievements make the Chinese people proud and the homeland beautiful. Some of the brightest technological and scientific Chinese abroad are heading back home. It is reversing the brain drain. A day before the launch, a Chinese spokesperson said, "It is an important advance for all of mankind in exploring outer space. China stands ready to cooperate with the rest of the world on the peaceful use of outer space."[16] Whether the goal is peaceful or it is for the development of space technology for military applications—for example, star wars technology, as the United States is pursuing—only time will tell. But technology cannot be confined to a single- purpose use when it could enhance national security as well. That explains why China is going the same route to space as Russians and Americans did.

National pride is very important. Such ventures may have diplomatic and political advantages, too. The Communist Party reinvigorates and reasserts itself through such achievements. It may have diplomatic impact upon China-Taiwan relations. Nuclear weapons, intercontinental ballistic missiles, and satellites give China all the trappings of an upcoming superpower. With double-digit percent annual economic growth for several years, China has the resources to pursue its space ambitions. It also makes you wonder how much individual freedom people need to accomplish so much in such a short time, but that is equally true of North Korea and Iran.

It is important, however, to see the Chinese space venture in a broader perspective, especially when the Russian space program has slowed down due to lack of funds and change of priorities, and the NASA space program is undergoing repurposing. Joseph Kahn of the *New York Times* might have summed up, albeit rather hastily, how the American elite feel about the Chinese launch: "The technology is vintage Kennedy and Khrushchev, the secrecy is reminiscent of the cold war and the popular reaction is a bit of fizzle, but China's seemingly successful effort to put a man into orbit is a sign that it intends to become a peer of the United States. The goals may still be 50 years away, but the direction is clear. Technologically and militarily, China is determined to retrace the steps that made the United States the world's unmatched superpower."[17] On June 24, 2012, a Chinese spacecraft carrying three astronauts docked manually with an orbiting module, a first for China as it endeavors to catch up with American and Russian achievements in space. To the genuine pride of the Chinese people, the *Shenzhou 9* capsule's maneuver with the *Tiangong 1* module was shown live on national television. The docking was carried out from a remote-controlled ground station.

CAN CHINA CHALLENGE AMERICA WITHOUT EMBRACING THE FIRST FREEDOMS?

Because of the absence of free speech, freedom of the press, fearless thinking, and accountability, there have been reports of continuing slave-labor work conditions and overwork deaths in China. Complaints of lead-paint-tainted children's toys, fake baby formula products, and sewage- and pollution-contaminated drinking sources of water in most major cities have been ceaseless.

Once a most egalitarian communist society, China has now become a most divided one with wide income disparities between the haves and the have-nots, which as Xubei Luo and Nong Zhu apologetically argue, "can be part of a normal process of development at a certain stage, and that the dynamics of spatial income divergence in the form of a race to the top can be desirable to some extent as it unleashes competitive pressure and creates incentives for investment in skills."[18] But the income gap is not shrinking if one considers China's Gini coefficient, "an income distribution gauge used by economists" that "worsened from below 0.3 a quarter-century ago to near 0.5 today."[19] Nonetheless, the Chinese believe that they are a unique people with a manifest destiny. Free press, free speech, free elections, and multiparty democracy are unnecessary. Doing independent research on topics such as the 1989 Tiananmen Square massacre and the Mao-era Cultural Revolution are off journalistic and academic limits. But at the same time restrictions on free inquiry and free speech leave the government blindsided about food and consumer product safety issues, especially when government regulations seem inadequate.

PRAGMATISM

Why would China censors—who exercise total control over human expression in any form including the news media, the Internet, and street protests—allow the press to rake up stories about food safety instead of letting the government food inspectors do their jobs and prosecute the criminals quietly? "Toxic bean sprouts, filthy cooking oil, drug-tainted pork: The relentless headlines in Chinese media have churned up queasy feelings for months about the dangers lurking in the nation's dinner bowls," according to a news agency report.[20]

The food safety problem is too big and hasn't diminished since the 2008 melamine scandal when six babies died and three hundred thousand fell sick after using the baby food formula and other dairy products contaminated with industrial chemicals. The recognition that the government's supervision of the food industry alone is not enough has prompted it to relax control over the press and let it gather public complaints and do investigative reporting for it. Letting the press do the "watchdog" function is much cheaper and more efficient than depending upon government inspectors who might be bribed or muzzled. Since the press freedom is not institutionalized and the Communist Party exercises complete control, the limited freedom can be withdrawn anytime. Perhaps there is a slow realization on the part of the Communist Party that the marketplace, whether it is food for domestic consumption or toys for export, needs the unbridled freedom of the press to scrutinize its claims. But the unanswered question is whether the barks of the "watchdog" can be limited to food safety only.

Zhiwu Chen, professor of finance at Yale School of Management, commenting on China's surging economy in spite of the global economic downturn, wrote in the *Financial Times* that although China is not known for market-friendly institutions, its economy has been growing at the rate of 9 percent a year. He argues that this is possible only because of China's export-oriented manufacturing, which does not require as much freedom of information as the service sector, which in China is poorly developed at present. Chen argues,

> Market development in manufactured goods is much less demanding than in services, especially financial services. The physical tangibility of a product leaves little room for cheating, giving such a market a natural informational advantage. In contrast, the intangible nature of "services" presents the buyer with severe informational disadvantage. In this case, both legal institutions and free media are critical. . . . *Press freedom facilitates service development by reducing information asymmetries among transacting parties, bringing confidence to the marketplace. Unbiased information enhances trust, which is fundamental to the deepening of a service market.*[21] (Italics added)

Being the global hub for manufacturing, mostly contract manufacturing, has taken a toll on the environment, which could have been prevented if the media were free. Moreover, Chen argues, China cannot depend upon export-oriented manufacturing for future growth to employ its migrant rural millions and therefore must develop its service sector. For a healthy service sector to develop, China will need institutional reforms, including "reliable law enforcement, judicial independence and freedom of the press." Freedom of the press will become a necessary condition for future market-based economic growth in China. Censorship is an economic bottleneck because the marketplace, especially service and financial markets, needs free, trustworthy, and useful information. Apart from supplying independent information to the marketplace, a free press also provides "a corrective mechanism for the economy in a way that the government cannot. Thus, media freedom is not only politically necessary, but also good for growth and job creation."[22]

DISSIDENTS IN THE DIGITAL AGE

In 2008 China was once again successful in crushing the Tibetan freedom uprising, which spread from the politically reorganized region known as the Tibet Autonomous Region (TAR) to the outlying provinces—Gansu, Sichuan, and Qinghai—where Tibetans have significant cultural and demographic presence. Once again, temporarily, Tibet was silenced and its nonviolent peaceful people withdrew into their hearts and prayers. But that could hardly be deemed as an extraordinary achievement for a rising global power, especially when it has been trying to show off to the world how the country's breathtaking economic growth has transformed the lives of the people, including, as China asserts, the Tibetans, who for centuries suffered the tyranny of isolation and serfdom.

As I discussed elsewhere, "During more than half a century of total domination over Tibet and its cultural and religious institutions, including the massive settlement of ethnic Chinese and their businesses into the heartland of Tibet, China was supposed to have reformed and re-educated Tibetans into total submission to the Chinese superior culture and hegemony."[23] So I wonder as to why China, one of the most ruthless authoritarian states mankind has ever known, has failed to indoctrinate and control the minds of a small minority of six million supposedly leaderless people? Why do these helpless, abandoned people listen to the unheard voice of the "splittest," persona non grata, the Dalai Lama, from the faraway hills of India, in Dharamsala? Didn't the mighty Soviet Union crush innumerable uprisings, including the bloodiest student-and worker-led brief revolution on October 23, 1956, in Hungary?[24] The Soviet Union sent tanks and troops into Hungary. In spite of the fact that there were worldwide protests against the Hungarian suppression, the Soviets succeeded in reestablishing their yoke on the country. One might argue that that was the era of Cold War, balance of terror, and mutual obliteration, and the United States and the Soviet Union had other conflicts to resolve. Hungary was shunted into the vast oblivion until the fall of the Berlin Wall and the breakdown of the Soviet Union in 1991.

HUMAN RIGHTS

But why is Tibet different? After the failure of the Tibetan uprising in 1959, the young Dalai Lama, along with eighty thousand Tibetans, escaped to India. And thanks to India's compassionate culture, the Nehru administration allowed the Tibetan leader to gradually establish an international presence, political and spiritual, instead of turning Tibetans into perpetual refugees. Since the first great escape, Tibetans, the young and the old, have been defying the hazards of the Chinese boots on the ground and have kept coming and settling in Dharamsala. Another 2,500 to 3,000 Tibetans will escape to India this year, and many of them will eventually go to Europe and the United States, keeping the spirit of freedom alive for Tibet.

The ascent of the Dalai Lama as a global spiritual leader is an incredible phenomenon. Nothing has shrunk him; nothing has diminished his stature. How could this unassuming monk, who has been called by a top Chinese official nothing but a "wolf wrapped in a habit, a monster with human face and animal's heart," whose people are undergoing a cultural genocide, still be so forgiving and so loving?[25] On her visit with the Dalai Lama in Dharamsala in 2008, former Speaker of the House Nancy Pelosi, a most courageous and compassionate American woman, told her audience that Chinese outrages against Tibetans were "a challenge to the conscience of the world." Indirectly perhaps, she was reproaching India for being too timid. "If freedom-loving people throughout the world do not speak out against China's oppression in China and Tibet, we have lost all moral authority to speak on behalf of human rights anywhere in the world," she lamented.[26]

In September 2007 German chancellor Angela Merkel received the Dalai Lama in her office as a goodwill gesture for a global spiritual leader. But Chinese authorities were furious because it amounted to giving the monk recognition as the political leader of Tibet. China-German relations had begun to improve until the peaceful demonstrations by Buddhist monks in Lhasa turned into a spontaneous upsurge in the entire greater Tibet region. Suddenly the massive public relations camouflage that the Chinese had succeeded in casting over the world about its improvement in human rights disappeared. The whole world watched what the Chinese authorities were doing to Tibetans, thanks to the liberating power of cell-phone video, social networks, and YouTube. But Tibetans have nothing to commemorate except remember their misery.[27]

Germans have been very sensitive about human rights issues because the genocide and Holocaust are eternally engraved into their consciousness and historical memories. "Violence can never be a solution. The two sides can only arrive at a solution through dialogue," German development minister Heidemarie Wieczorek-Zeul was quoted in *Deutsche Welle*.[28] But how can there be any dialogue between two sides unless they accept each other? Zhang Qingli, Tibet's Communist Party secretary, told regional officials: "We are engaged in a fierce battle of blood and fire with the Dalai clique, a life-and-death struggle between the foe and us."[29] The world is waiting for China's "peaceful rise" and the "journey of harmony"! Germany was outspoken about China's oppression in Tibet, and even suspended environmental technology aid talks. Former British prime minister Gordon Brown was no less audacious in his decision to meet with the Dalai Lama when he visited London. Brown briefed the parliament about his phone conversation with Chinese prime minister Wen Jiabao, admonishing him that "there had to be an end to violence in Tibet . . . an end to the violence by dialogue between the different parties."[30] China deems the Dalai Lama as an enemy of the state rather than a party to the dispute.

China could resolutely crush the Tibetan people's uprising, as the Communist Party's newspaper *People's Daily*, advocated to turn Tibet into an Orwellian Panopticon. The Soviet Union had done so in Hungary and its other satellite countries, but those were different times. This is the age of wireless mobility, texting, networking, and decentralized global organizations like WikiLeaks exposing corruption and human rights violations everywhere. Can the Chinese Communist Party succeed in using the ubiquitous Chinese propaganda machine and the digital firewall to control the Tibetan mind without doing the same to the rest of the population? Can a fearful society become a global superpower in the digital age?

Virtual Struggle

The continuing struggle in Tibet demonstrates how new web technologies, cell phones, digital photography, texting, and email, are making it harder for Chinese communist authorities to control the struggle for self-rule and freedom.[31] Global networks are shrinking space and time, transforming geographic space into cyberspace, and are bringing people together for collaboration in the international political space. The whole world has been experiencing the agony of Tibet in a virtual environment created by global-networked communication since the renewed struggle began in early March 2008 with Buddhist monks' demonstration in Lhasa. Web technologies help us improve our awareness and responsiveness to the human condition in remote corners of the world. From Tibetan monks crying in public, "We have no freedom, we have no freedom," to the most complex genomic problem, every human activity is nothing but a stream of information pouring into Big Data. Whatever takes place in the real world can be turned into a digital stream and instantly distributed globally through networks, thus extending the reach of humans and empowering them.

Images can be stored, retrieved, and replayed and recontextualized for any purpose anywhere. Major newspapers not only archive printed stories but also videos of the ongoing struggle around the globe. The same information can also be transformed into intelligence about human behavior regarding commerce, national security, or any other social or political activities. Political discourse, slogans, and chants and cries for freedom become indistinguishable as they converge in a digital flow and surge through cyberspace. Today the Chinese people may be prevented from seeing Tibet sufferings on YouTube, but the rest of the world will be playing it again and again. If a picture is worth ten thousand words, as Confucius might have said, YouTube is forever, as Chinese, too, will discover one day.

Spontaneity, instantaneity, mobility, and convergence make the Internet the most powerful medium of resistance ever developed.[32] Since the traditional media including books, television, newspapers, magazines, radio, music, and interpersonal communication are converging on the Internet as a multimedia stream into which anyone can plug in, their power increases manifold and in ways whose implications Chinese authorities fear tremendously. YouTube, for example, has given Tibet a constant state of telepresence in the European and American political discourse.[33] Normally we talk of off-shoring of manufacturing and outsourcing of research and development and other forms of intellectual and professional work to other countries. But China, because of its paranoid nationalism, has been off-shoring the Tibetan people's struggle and Buddhism to Europe and the United States, thus globalizing the Tibetan issue. The Internet thus is revolutionary in the sense that it is lowering barriers for cross-border convergence of cultures and human rights activism, which China cannot escape for long.

Although people in ancient times endeavored to bridge space and time by sending messages, especially in wartime through drums and smoke signals, it was not until the invention of the telegraph that it was possible to think about communication in terms of transmission rather than transportation. Like goods, messages were communicated from place to place at a speed that the best transportation system of the time, for example, the camel, the donkey, the pony express, or later on the railroad, made possible. The telegraph changed the geography-based metaphor of communication, which was no longer synonymous with physical transportation. As the telegraph prompted the development of new technologies in the early part of the twentieth century, telephone, radio, and television became omnipresent, and communication became increasingly released from space-time constraints. Computer networks and the Internet have further altered our view of space-time. A networked organization or an individual with texting and instant messaging has a different feel of space-time than the people of the predigital era. Mobile phones are the door to cyberspace, and once you are there, you are simultaneously in a synchronous and asynchronous world, a world that gives a greater sense of freedom and control than the real world. That's how international organizations in the forefront of Tibetan liberation are helping Tibetan people to continue their resistance and stop the Chinese authorities from turning Tibet into an American-style native reservation.

Cyber Resistance

The Chinese dilemma is that as more and more people experience Tibetan resistance in cybermedia, the authorities don't know what to do about it. Under the advice of public relations firms, China conducted a controlled tour for journalists, but that did not help to "reeducate" the untamed minds and hearts of the Tibetan people. The enemy is everywhere in cyberspace. The Canadian scholar Herald Innis said in *The Bias of Communication* that a new medium of communication creates a specific cultural shift and changes our concept of space and time, with tremendous cultural consequences.[34]

The Chinese public relations offensive included inviting a handful of selected diplomats on an officially controlled trip, a visit that came after the foreign journalists on a tour of Lhasa encountered a group of thirty Buddhist monks shouting that there's no freedom in Tibet. The Han Chinese "demographic aggression" into Tibet, which the Dalai Lama referred to, was glaringly apparent during the monks-led anti-Chinese demonstration that began on the forty-ninth anniversary of the Tibetan uprising against Chinese occupation.[35] Besides its crude public relations propaganda and diplomatic intimidations, China is also fighting a cybercampaign with its Operation Golden Shield, the so-called Great Firewall of China. Of course, technologies of freedom can also be used for repression, especially inside China. But there is a cyberworld beyond China, over which no one has any control.

THE PERSISTENCE OF CHINA'S DISSENTERS

Like the Chinese dissident and democracy advocate Liu Xiaobo, who on Christmas Day 2009 was jailed for eleven years for "incitement to subvert state power" and was later awarded the 2010 Nobel Peace Prize, many Chinese intellectuals must wonder, perhaps like T. S. Eliot's J. Alfred Prufrock, "Do I dare?" "Do I dare disturb the universe?"

Most Chinese would rather make money than risk their lives for political freedom. But those who dare must keep in mind that anyone with seditious and insidious ideas who disturbs social harmony and the peaceful rise of China as a dominant global superpower must be silenced. China's one-party authoritarian leadership has little tolerance for the prolonged grueling primaries and presidential elections and gridlocks on which American democracy thrives. Aid activist and environmentalist Hu Jia was sent to prison for three-and-a-half years for his subversive writing. In 2005 another writer, Shi Tao, was convicted and sentenced to ten years for leaking to an outside non-Chinese website a Communist Party internal memo.[36] The blind civil rights activist Chen Guangcheng, who received international recognition for fighting for the rights of rural people, had to seek refuge in the U.S. embassy after he escaped his pursuers in April 2012. He is now in the United States hoping against hope that he will return to China, or perhaps Americans will take up his cause. Americans would rather do business with China than help freedom lovers, if American attitudes toward Tibet are any indication.

According to the Committee to Protect Journalists, out of the 179 journalists jailed worldwide in 2011, twenty-seven were jailed in China.[37] And foreigners must behave properly if they want to do business with ever-growing China or borrow money from its trillion-dollar reserve funds to sustain their weak-legged economies or want to use its diplomatic leverage with nuclear North Korea and other rogue states such as Iran, or have plans to do something about global climate change. China says it can do all this and much more for you and, in fact, we have become so interdependent that whatever China does or does not do at home or abroad impacts the rest of the world. Many countries in Asia and Africa envy China and would like to emulate its freedom-neutral, export-based model of rapid economic growth.[38]

In spite of its miraculous economic growth, China is terribly afraid of upsetting its delicate apple cart, afraid of the power of ideas of men like Liu who refuse to bend and bow before the might of the one-party authoritarian rule. It has been much easier for China to condemn the Dalai Lama, a "splitter" of the motherland and a "devil in a monk's robe," but what to do with Liu Xiaobo, who at the time of 1989 Tiananmen Square massacre was a visiting scholar at Columbia University but returned posthaste to China to participate in the protest, for which he had to spend twenty months in prison.[39]

It must be a profound puzzlement for the Communist Party bosses to understand why Liu Xiaobo still talks of human rights and democracy while China has been going up from economic peak to economic peak to dazzling Olympian heights. For his persistent writing about sociopolitical conditions in China and his contribution to Charter 08, Liu was held in custody for more than a year before a formal indictment.

Charter 08 asks for fundamental rights, the rights to which people in Europe, the United States, India, and other liberal democracies have become so accustomed that they seem so natural. Charter 08 includes rights such as: separation of legislative, judicial, and executive powers; legislative democracy; an independent judiciary; freedom to assemble and form groups; freedom of expression and religion; civic education; protection of private property; and most of all, a federated republic.[40] Ten thousand Chinese who are reported to have signed the document since its inception on December 18, 2008, form only a tip of the iceberg. In December 2009 Liu was subjected to a hasty trial from which not only his wife was excluded but also the news media. Liu, who has received international accolades for his fearless human rights advocacy in China, won't be allowed to write anything during his prison term because under the current Chinese law he will lose all his political rights.

International protests do not matter to Chinese authoritarian rulers who are determined to maintain their hold on power in the name of social harmony and economic growth. Since Liu's arrest in early January 2009, several organizations, including Human Rights Watch, Amnesty International, Reporters without Borders, PEN, as well as many prominent writers (Salman Rushdie among them) and Nobel laureates have appealed to Chinese president Hu Jintao, but to no avail.

President Obama might have timidly raised the issue during his Beijing visit in 2009, but having declined to meet with the Dalai Lama, Obama's commitment to human rights in China seems rather dubious in comparison with his predecessor George W. Bush's call for democracy and human rights.[41] Former Secretary of State Hillary Clinton, thanks to her newfound political pragmatism during her tenure, seldom talked about Chinese human rights, though as the First Lady when she visited Beijing in 1995 to attend the United Nations Fourth World Conference on Women, she forthrightly told her Chinese audience, "Freedom means the right of people to assemble, organize, and debate openly. It means respecting the views of those who may disagree with the views of their governments."[42] Liu Xiaobo and the Charter 08 signatories are asking for the same thing; they won't go away. The biggest challenge to China's one-party authoritarian rule comes not from its peripheral regions, Tibet or Xinjiang-Uygur, but from its heartland, from people like Liu Xiaobo, who at the time of being awarded the Nobel Peace Prize in absentia, said in "My Final Statement":

> I have no enemies, and no hatred. . . . For hatred is corrosive of a person's wisdom and conscience; the mentality of enmity can poison a nation's spirit, instigate brutal life and death struggles, destroy a society's tolerance and humanity, and block a nation's progress to freedom and democracy. I hope therefore to be able to transcend my personal vicissitudes in understanding the development of the state and changes in society, to counter the hostility of the regime with the best of intentions, and defuse hate with love . . . I do not feel guilty for following my constitutional right to freedom of expression, for ful-

> filling my social responsibility as a Chinese citizen. Even if accused of it, I would have no complaints. [43]

WILL INFORMATION BE FREE IN CHINA?

As China's economy began to change from a communist-style, absolute state control to a state-controlled capitalistic economy, similar changes were also effected in China's mass media companies. According to Susan L. Shirk, although since 1979 mass media companies have been allowed to compete in the marketplace for advertising revenue, they remain under the control of China's Communist Party (CCP) for editorial guidance.[44] Competing in the marketplace for audience has jazzed up the Chinese media and made newspapers, magazines, and television much more interesting for consumers than they were earlier, though CCP's total media control remains unchanged.

But the Internet and the proliferation of new social media have made life difficult for CCP. Rising incomes, accessibility to various sources of news, information, and entertainment, and global exposure through the Internet has wetted the Chinese people's appetite for authentic and reliable sources of information. The CCP has been using the market-oriented, traditional mass media as a surveillance tool to gauge emerging trends and to control them before they become a serious threat to the Communist Party rule; but the Internet is a different medium.

The Internet's many-to-many capabilities, simultaneity and speed, and decentralized nature make it difficult to exercise complete control, in spite of the notorious fifty-cent army of spin-doctors.[45] China's 2012 Internet population of 538 million, including 300 million microbloggers who use a Twitter-like service called Sina Weibo, is potentially an explosive force that can bypass the state-controlled traditional media and seek information wherever available, in spite of the great Firewall of China, especially in the age of convergent media, when hot news jumps from platform to platform and assumes many formats and disguises. The sheer abundance of Chinese media, newspapers, magazines, television programs, and the mobile Web presents the greatest challenge to Chinese censors.[46]

Chinese websites, even though under state surveillance, aggregate news and opinions from all sources, including foreign media, so long as they do not question the CCP and the official line of thinking. The publication of local corruptions and disasters and scandals abroad do not undermine the authority of CCP; nonetheless, such stories give the people an illusion of freedom of the press, even though Freedom House lists China at the bottom of the free speech ranking. China's self-cultivated global image of a harmonious society on a high-speed escalator of economic growth is more important than free speech, and hence the exercise of extreme control over media during the 2008 Olympics and anniversaries such as the Tiananmen Square Massacre, and its struggle with Internet companies, especially Google.

Google is an information company, not a contract manufacturer, and its business model is predicated upon the free flow of information. Having initially succumbed to pressures from the Chinese government to filter out sensitive information that's inimical to the state policies as a price for doing business, Google realized it could take censorship no more and moved out of the country, especially when it found that Chinese authorities had been tolerating hacker attacks on Google's email.[47] If Google's struggle with China symbolizes Internet freedom as an American value and a foreign policy goal, any compromise would have been unlikely.

In fact, Internet freedom is the greatest challenge the Chinese Communist Party has ever faced, a challenge that CCP is unlikely to survive because at some point unfettered freedom of speech and unhindered flow of information will become an indispensable factor in economic growth and global trade.[48] Besides, as Shirk says, the commercialization of mass media in China has resulted in the loss of some control, which inevitably "transforms the strategic interaction between rulers and the public in authoritarian political systems like China."[49] Free flow of information from the public via the media can enable the market to work more efficiently and help authorities to recognize problems and solve them before they become too serious to handle. But what is the optimum level of press freedom that creates enough transparency to enable efficient governance without challenging the authoritarian rule? Freedom of the press is not a halfway house, and relinquishing control over information engenders tremendous political risks for which the Communist Party, used to exercising unquestioned authority, is not well equipped to handle except through brute force and nationalistic propaganda, as has been evidenced repeatedly in Tibet, Xinjiang, and Inner Mongolia.

Abundance without Freedom

China has 2,200 daily newspapers, some of them published from multiple centers, and more than eight thousand magazines and journals, besides approximately seven hundred television stations and three thousand cable channels as well as one thousand radio stations. Most of them are advertisement supported, but none is editorially free; if someone tries to break the CPP-imposed editorial shackles, the punishment, apart from losing one's job, can be severe. Unsurprisingly, Chinese journalists are easily prone to corruption by corporations and businesses.[50]

In the age of the Internet with its decentralized freedom, spontaneous and instantaneous twittering and social networking, free-wheeling bloggers, and the medium's drumbeating redundancy, the challenge for the Chinese authorities is how to use new media for collective mind control before millions of dispersed voices converge into flash mobs to oppose the government or congeal into a political opposition.[51] Consequently, the Chinese authorities are using all the available means—technological, social, and punitive—to control online content and prevent it from becoming a platform for virtual activism that might eventually spill into the streets and become uncontainable, as it has become in Middle East countries.

Will the lapdog become a ferocious watchdog? So long as the CCP leadership remains united and has the full support of the military, it is very unlikely that a challenge on the scale of Tiananmen Square will erupt again. But commercialized media seeking audiences and advertising support from the open market might become subversive in thousands of small and unobtrusive ways until the CCP loses public backing. Instead of a big, spectacular conflagration that might draw global attention, imagine the Chinese media, egged on by millions of bloggers and microbloggers, reporting daily about millions of small fires, all over the county—daily protests against local crimes and corruption, police atrocities against unauthorized migrant workers, factory accidents, defective products and services—all quite legitimate activities of the press in a free society. They might become acceptable in China as normal, but eventually millions of micro-mutinies will dissipate the one-party rule even if the authorities continue taking corrective measures in response to diffused media criticism.

This new dynamic between the one-party authoritarian government and market-oriented media struggling for greater freedom, especially when authorities want to use the media for gauging public opinion and surveillance functions, will have tremendous consequences for China. Internal divisions in the party leadership and among various regions and ethnic groups will gradually get media exposure. This is what is happening to Chinese businesses, whether it is shoddy constructions, abuse of migrant workers, or melamine-tainted milk or infant formula. At present Chinese authorities give the news media enough freedom to know what is happening in the country, but once a problem is known, they clamp censorship through top media editors or through the fifty-cent army of progovernment Internet commentators.

If the Chinese people have to choose between the official media outlets like the party newspaper *The People's Daily* and professionally produced, albeit semifree commercial media, they are likely to choose the latter along with proliferating Internet media. The search for authentic information is a universal impulse, especially during crisis, whether it is a natural disaster like an earthquake, an outbreak of an epidemic like the severe acute respiratory syndrome (SARS), or public, official, criminal negligence regarding food safety.

Government transparency through the media might become a necessary condition for the CCP rule to survive, but on the other hand, the very transparency might spring forth alternatives to public policy, further diluting absolute political power. Even unbridled access to news and commentaries about sports and entertainment, popular culture, science and technology, or what is happening abroad creates a desire to know more about what is affecting people's daily lives.

HALF-HEARTED FREEDOM

The question is whether a half-free society can ever challenge American exceptional powers that grow out of First Amendment freedoms. Rising living standards increasingly make the Chinese aware of their politically oppressed conditions, especially when they compare themselves with open societies in the West. Eventually they are bound to chaff at multilayered government controls.

At the time of launching his White House bid as a Republican candidate in June 2011, former U.S. Ambassador to China, Jon Huntsman, said that growing tension between the rising economic class and the ruling political class did not bode well for China's peaceful rise and domestic harmony. Huntsman said,

> And at some point, as you get 300 and 400 million people who now become members of what will be the largest middle class in the world, they're going to have to make some really tough choices about how you bridge the economic and the political chasm, which is not easily done . . . If they don't do it, they're likely to have some sort of head-on collision, I would say four or five years into the future. . . . If you do not have political freedoms that run in parallel with economic development and reform—if you don't have political reform—at some point you're going to find yourself in a pretty precarious situation, one that will force you to address it as opposed to allowing you to evolve.[52]

The extraordinary single-minded focus by the CCP's nine-member Political Bureau on transforming China into a global economic and military superpower can be maintained only through extreme nationalism and reassertion of party control. But in spite of rapid economic growth, every year there are thousands of protests, riots, and strikes and even suicides in different parts of China. And they have one common denominator of grievances: a helpless sense of injustice. Millions of migrant workers are treated as second-class Chinese citizens by the household registration system, which deprives them of equal access to health care and education.

In Zengcheng, Guangdong province, the industrial heartland of China, thousands of workers protested for three days in June 2011 after security forces assaulted a pregnant woman. Elsewhere, migrant workers protested against unpaid wages and lead poisoning. In Inner Mongolia, when ethnic

Mongolians protested against coal mining companies destroying their culture and polluting the environment, ethnic Han mineworkers turned violent and killed many ethnic Mongolians.

Authorities have been responding to protests with brutal suppression to restore order, a short-term tactic that seems to have worked so far. Land grab by local authorities and demolition of old houses for commercial development without proper compensation or the right of appeal is so common that ordinary Chinese feel totally helpless and sometimes resort to violent protests.[53] On May 26, 2011, in Jiangxi province, Qian Mingqi, whose property was confiscated without proper compensation, targeted government buildings by setting off three car bombs that killed four people, including him. On June 4, 2011, in Hubei province more than 1,500 people besieged a government building to protest against the torture and death of a local legislator, Ron Jianxin, a fighter against corruption, who was accused of bribery and died under the police brute interrogation.

This is not the picture of a harmonious society aspiring to become a global superpower and a growth model for the world.[54] As Max Fisher wrote in *The Atlantic* that Chinese authoritarian government's great unspoken bargain with the Chinese people to submit to its total control in exchange for a steady and high-rate of economic growth is losing its attractiveness because of the "arbitrary rule, lack of political participation, coercion . . . And though China's unrest appears unlinked to the uprisings in the Arab world, they are both enabled by similar factors: the rise of social media as a tool to organize and to broadcast images of government abuse, the influx of liberal democratic ideas, and a worldwide opening that shows Chinese and Arab subjects alike that life doesn't have to be this way."[55]

THE FUTURE IS A DRAGON. DO YOU HEAR IT COMING?

The Chinese dilemma is profound. It wants to be an innovative society without free speech and freedom of the press. "As a national strategy," according to author and journalist Steve Lohr, "China is trying to build an economy that relies on innovation rather than imitation. Clearly, its leaders recognize that being the world's low-cost workshop for assembling the breakthrough products designed elsewhere—think iPads and a host of other high-tech goods—has its limits."[56] But how much innovation can take place in a society where the clash of ideas in open forums is not tolerated because of the fear that ideas might cross boundaries with unintended consequences? Even an incremental innovation can be a challenge to authorities. John Kao, an innovation scholar, says in his book *Innovation Nation*, "American culture, more than any other, forgives failure, tolerates risk and embraces uncertainty. . . . But America's future lies in being the orchestrator—the systems integrator—of the innovation process. Look at Silicon Valley. It is a place where smart people from all nations, all languages and all ethnic groups come together. It's the capital of innovation assembly."[57]

The inflow of the best brains from all over the world into Silicon Valley is predicated upon the culture nurtured by First Amendment freedoms that create vast spaces for people to let their imagination go wild, to think outrageously, tinker, expand, experiment, compete, rise, fall, and rise again. Silicon Valley without the First Amendment will be like the Gobi Desert.

It has been much easier for the Chinese censors to control the traditional commercialized media through coercive persuasion, outright intimidation, and getting top editors fired and putting putting a choke on advertisement. In mass media organizations there are some regulatory buttons and gatekeepers, which can be easily controlled.

Even in the United States the government and powerful corporations sometimes succeed in killing or delaying controversial stories. In China, media business is no different from other businesses that are by and large under state supervision. The traditional media's freedom is limited to making money by maximizing their audiences without questioning the CCP control and government policies. Getting rich without challenging the authoritarian party rule has been working well for the Chinese media and fits into the overall pattern of state-directed capitalism and total party control. The implicit understanding is that to make money you need freedom to compete in the marketplace, but not freedom of speech. A person can be free to work without much free speech, as has been the case with a most famous Chinese artist, satirist, and government critic Ai Weiwei, who was arrested not for challenging the government but for tax evasion.[58] After three months of remaining in jail, he was released with a gag order on his head, but he continues working by walking a fine line between defiance and obedience. One might say the Chinese traditional media are free without free speech. Freedom without free speech is a great Chinese political innovation. But every innovation has its shelf life.

POINTS OF LIGHT

In the age of the Internet there are millions and millions of dispersed and decentralized points of light without a central switch. And that has presented the greatest challenge to the absolute command and control of the Chinese Communist Party that has been governing the country by turning it into one massive, integrated intelligent grid. Working smoothly and silently, the grid attracted enormous amounts of direct foreign investment, transforming China not only into a colossal manufacturing hub but also a center for imitation and emulation of imported technology. The miraculous transformation of China into the world's second biggest economy in a generation has catapulted three hundred to four hundred million people into the upwardly middle class. These people feel more comfortable in the ecology of new social media, which makes them consumers and creators of information in a sharing and collaborative environment rather than recipients of remotely controlled, official, albeit market-oriented, well-produced, glossy traditional print and broadcast media. The rise of millions of Chinese netizens is one of the greatest phenomena of recent times, with a potential for releasing enormous entrepreneurial energies for commerce, trade, and innovation, if given unrestrained freedom. Also, this kind of dense connectivity generates demands for political reforms, which may be inimical to the state control over information.

Certainly the Chinese Communist Party is not going to give up its total control over society and has been using various methods to control the free flow of information enabled by the Internet. In an illuminating essay on online censorship in China, Xiao Qiang, who teaches at the University of California at Berkley Graduate School of Journalism, wrote that the government has been using "a multilayered strategy to control online content and monitor online activities" by using "a complex web of regulations, surveillance, imprisonment, propaganda, and the blockade of hundreds of thousands of international Web sites at the national gateway level, known as the Great Firewall."[59] The firewall consists of: keyword filtering that blocks content that the government considers might threaten stability such as Tibet, Falun Gong, dictatorship, and the Jasmine Revolution; service and access providers' liability for their customers' behavior; and tens of thousands of human monitors who read and censor objectionable content. Besides, China's nine Internet gateways block thousands of objectionable foreign websites. The major goal of the government's censorship is to prevent the formation of mass action through public demonstration, especially in big cities where apart from local and foreign media, ubiquitous cell phones and bloggers can spread information instantly, causing tremendous embarrassment to the authorities. But how effective is the control over the Internet?

INNOVATION IN A CLOSED SOCIETY

Qiang believes that the widespread Internet censorship and surveillance causes tremendous resentment among the Chinese netizens, provoking them to ridicule authorities and their sloganeering about the pretense of building of a harmonious society.[60] They mock authorities through "political satire, jokes, songs, popular poetry, code words, mockery, and euphemism," which is gradually diminishing the hold of the party on people's imagination and its agenda-setting power.[61] Whatever is censored in China is picked up by overseas Chinese bloggers and news media, which is elaborated and recycled and redistributed to China by information-savvy netizens who know how to bypass the Great Firewall.

Through these indirect and clandestine means, the information might spread like wildfire depending upon how crucial the information is before the party's propaganda department can reimpose its control. China's growing online population and the party's propaganda department continue playing the cat-and-mouse information game of censorship, exposure through ridicule, and regurgitating of blockaded information through foreign sources. In this information-disinformation game, journalists, according to Qiang, play a double game; if they cannot publish a sensitive story in the traditional media, they do so online, which then the traditional media pick up as something too important to ignore, and thus the information finally reaches the government and the masses, albeit in a circuitous ways.

This is not the picture of a totally unfree society; rather, an emerging society where authorities do not know what to do with the flood of information unleashed by new information technologies' multiple platforms.

But changes are occurring in spite of the tight control over media content by the Central Propaganda Department and other government authorities. For example, the Chinese news media initially reported freely and critically the July 23, 2011, two high-speed train collision in Zhejiang province that killed at least thirty-eight people and injured 192, though subsequently authorities clamped down on censorship and asked reporters and television stations to report constructive stories. Some journalists who were defiant lost their jobs. As Qian Gang, director of the China Media Project at the University of Hong Kong, wrote in the *Wall Street Journal*: "This is today's China: rapid development attended by deep crisis, but with citizens who are gradually awakening. With each event, the media and the people resist and struggle, inch by inch, expanding their right to know, their right to participate, their right to express and their right to monitor."[62]

The Internet and microblogging played a big role in spreading the information "right after the crash at least 40 minutes before the first report from the official Xinhua news agency. From that point on, microblogs were the most effective media in urgently pushing for the truth, raising questions, pulling people together and mobilizing them. While the authorities would like to control microblog platforms, this medium is more difficult to control than, for example, television."[63]

Qiang quotes a journalist's microblog whose poignant words echoed in cyberspace: "China, please slow your soaring strides, wait for your people, wait for your soul, wait for your morals, wait for your conscience! We don't want derailed trains, collapsing bridges, roads that slide into pits or homes that collapse. Move more slowly. Let every life have freedom and dignity.

Let no one be cast aside by this age. Let every person reach their destination easily and in peace."[64]

CHALLENGE TO THE UNITED STATES

Does China have the unique cultural power to challenge the United States and the West? The fear of unbridled digital freedom and free flow of information from global networks has created a siege mentality, compelling the government to use vast resources to control the ears and eyeballs as well as the mouths of its people so that they don't challenge its authority. Not only do the Chinese people need mind control but also physical control as well. Andrew J. Nathan wrote in *Foreign Affairs* that to preserve domestic peace and harmony, China has to deploy massive resources "to maintaining control over two-fifths of its territory that comprises Xinjiang and greater Tibet, to keeping civil order throughout the densely populated and socially unstable Han heartland, and to deterring Taiwan's independence."[65]

What has kept the Chinese political and economic model viable so far is the vested interest of global corporations to use China's docile and cheap but hardworking workforce for contract manufacturing. Apple Corporation, for example, is more interested in the smooth running of the Chinese supply-chain system that assembles its iPad and iPhone rather than free speech and freedom of the press that might howl against the awful working conditions at Foxconn that have led to many suicides. But that is not the case with Google, who now is as much interested in uncensored free flow of information and protection of intellectual property as is the rest of corporate America. Some Americans think that a long-term strategy for dealing with China should be to enable the Arab Spring to spring up in China.

China's authoritarian one-party rule will dissipate into an open and transparent multiparty system for which the Internet and new mobile media technologies are the greatest means of liberation. But that would take a while. In the meantime, we should keep this fact in mind: Authoritarian China cannot handle political freedom because it does not have the American cultural capabilities to do so, without which it cannot establish its supremacy and challenge America even if it continues growing at the present rate, which is more than unlikely.

The source of the unique American cultural capabilities is the First Amendment freedoms, which China cannot import without transforming itself into a different society, one that will closely resemble the United States, the aspirational society that has taken decades and decades of trial and error to arrive at a system that maintains a dynamic equilibrium between freedom and control to break the decline-and-fall pattern of history.

Chapter Seven

Political Innovations of the Aspirational Society

TO LIVE AND DIE IN FREEDOM; THAT'S AMERICA, BUT WHAT ABOUT THE REST OF THE WORLD?

New York is the heart of American civilization. It is the city of extreme aspirations. If Al-Qaeda terrorists aimed to create widespread panic among the American people and paralyze the U.S. government, they were a terrible failure—in spite of their daring and spectacular success in ramming two hijacked civilian jets into the soaring twin towers of the World Trade Center, another into the mighty Pentagon, and crashing the fourth into a field in Pennsylvania on September 11, 2001.

A new World Trade Center complex with the National September 11 Memorial & Museum has been rising below and over "Ground Zero," floor by floor as a symbol of America's will and capacity for rebirth and self-renewal.[1] And millions of people from all over the world keep coming to New York, the most innovative, colorful, and vibrant place on earth—the city of wealth, art, and culture.

Sooner than one could have imagined, New Yorkers were up and working to heal the wounded and recover the dead from the steel and concrete rubbles, and to build again. Rudolph Giuliani, who was the city's mayor, looked like the bravest of the brave New Yorkers, fearlessly walking the streets of the city waving his arms, spreading solace and hope, moments after the Twin Towers in lower Manhattan had come crashing down.[2] It was Giuliani's finest hour. Giuliani stepped down as the mayor of New York—the city that under his leadership became a place to live and love again and in the process healed the psyche of New York.

The terrorists' well-coordinated simultaneous attacks and ingenuity in using domestic air flights as instruments of death and destruction turned the initial shock and disbelief into a grim, warlike mood, not only to retaliate but also to punish—to vanquish the enemy forever as Americans did during World War II. Americans not only crushed Japan and Germany, they also turned them into civilized trading partners in global coprosperity. Germany and Japan will never be a military threat again. That was the steely resolve after 9/11—cold rage against terrorism, the international menace that must be eradicated.[3]

The Bush administration was utterly determined to meet the challenge, whatever and however long it took, though the enemy was radically different—a decentralized, loosely networked organization of proliferating deadly cells, sheltered and nurtured in faceless anonymity, and quietly tolerated and in some cases encouraged by host countries. Freedom was under attack, and both the perpetrators and the collaborators must be dealt with severely, said President George W. Bush during the address to the nation in the aftermath of the attack. Although much disparaged now, historians might regard Bush's words at the moment of diffused and widespread fear as inspiring and aspirational:

> A great people has been moved to defend a great nation. Terrorist attacks can shake the foundations of our biggest buildings, but they cannot touch the foundation of America. These acts shatter steel, but they cannot dent the steel of American resolve. America was targeted for attack because we're the brightest beacon for freedom and opportunity in the world. And no one will keep that light from shining. . . . America has stood down enemies before, and we will do so this time. None of us will ever forget this day, yet we go forward to defend freedom and all that is good and just in our world.[4]

Americans had never been so united before. Nor did I understand the meaning of "One Nation under God" as I did when I saw Americans fall on their knees, pray, and sing, "God Bless America." Pearl Harbor was far away, at the periphery of historical imagination. This time it was not merely a blind rage that would be easily quenched by lobbing a few missiles here and there, as President Bill Clinton did in 1998 after the bombing attacks on U.S. embassies in Kenya and Tanzania.[5] Those half-hearted measures against terrorists did not work. What was required was nothing short of a systematic and total destruction of the entire network of terrorism, if Americans wanted to preserve a free and open society and the principles they stand for. But freedom is not the preserve and the aspiration of only the United States.

Nor was the United States the only country that suffered terrorists' attacks, and soon it began to dawn upon terrorism strategists that the United States could not fight terrorism alone. If it's going to be a war against terrorism, there must be a war strategy and battle plans, and America needed the cooperation of NATO allies and other partners in peace, especially South Asia, where Pakistan-supported militants carried out daily acts of terrorism against innocent civilians. NATO activated Article 5, under which an attack on a member is an attack on the whole alliance, requiring members to open all facilities for retaliation against the enemy. Saudi Arabia, Kuwait, and Egypt could not survive a day without U.S. support.

In less than nine months into his presidency, President Bush abandoned his unwise and ineffectual policy of withdrawal and semi-isolation from the troubles of the world. It was time to test him as the commander-in-chief of the greatest military and economic power, the leader of the sole superpower committed to freedom and democracy. He had to share America's resources in intelligence gathering, diplomacy, and military technology to fight terrorism, which must be fought at several levels. President Bush had to show the way to fight and win the war against the worst plague that afflicted humanity without compromising too much with individual liberties in spite of what euphemistically came to be known as the USA PATRIOT Act. The acronym shows the American penchant for linguistic inventiveness and stands for Uniting (and) Strengthening America (by) Providing Appropriate Tools Required (to) Intercept (and) Obstruct Terrorism Act of 2001.

The PATRIOT Act substantially reduced constraints on law enforcement agencies' abilities to gather intelligence within the United States. The Act increased the Secretary of the Treasury's authority to regulate financial transactions, especially those involving foreign individuals and nonprofit organizations. Under the Act, law enforcement and immigration authorities had greater discretion in dealing with immigrants suspected of terrorism-related acts. The Act eliminated the distinction between domestic and foreign acts of terrorism, thus expanding the law enforcement power to reconnoiter Americans.

On May 27, 2011, President Barack Obama, who was in France attending an international summit, signed a four-year extension of the key provisions in the PATRIOT Act, and soon after Congress, bringing together liberal Democrats and conservative Republicans, passed the surveillance law. The occupants of the White House change but the system continues. All shades of gray look the same from a historical distance.

THE DECIDER-WARRIOR PRESIDENT

Americans are a pragmatic, down-to-earth people who prefer leaders that can talk to them at their own level. This is especially true when they feel threatened at home and abroad, when their foes are cocooned in sleeper cells, corporate boardrooms, the anonymity of the streets, and even places of worship. In times like these, Americans see things in white and black, good and evil, us and them. And they looked at President Bush as the epitome of a very determined American leader, a kind of hard-nosed warrior president, someone with integrity, a role President Bush played well and exulted in his achievements. In an interview with Peter Robinson of the Hoover Institution, he said:

> Eight years was awesome, and I was famous and I was powerful. But I have no desire for fame and power anymore . . . I believe we are a blessed nation; that is, therefore, we have a sense of responsibility to the extent we can to help others. In this case there was a pandemic destroying an entire generation of people. And I didn't see how I could be president of a powerful, the most powerful, and the richest nation and not lend our support to saving lives. It would have been unconscionable not to act. So I thought it was in our moral interest to act. I also knew it was in our national security interest to act.[6]

President Bush could not have worked except in this warrior-savior mode because it helped him not only to rally his own energies, as we saw in the immediate aftermath of 9/11, but also to energize others to the national cause as he envisioned it. The warlike mode of thinking and doing things also enabled him to turn his weaknesses into strengths; for example, it let him to overcome his linguistic infelicity, and so he spoke in simple, rhythmic sentences that fell like a hammer. Consider this when President Bush conjured up the image of Saddam Hussein with nuclear weapons and why he must be disarmed: "He's a threat to the country. He's a threat to the people in his neighborhood. He's a real threat, and it's now time for the world to come together and disarm him."[7] There was no sophistry, subtlety, ambiguity, or multiple shades of gray in these utterances, and they left no scope for secondary interpretation. And he went on repeating the same theme until most of the American people began to believe in him.

His simple, direct message coupled with the readiness to use force, and some diplomatic compromises, turned around the UN Security Council to vote fifteen-to-zero in 2002, which included even an Arab country like Syria, demanding that Iraq must disarm.[8] This was a most remarkable diplomatic achievement, especially when it happened immediately after the 2002 midterm election results when President Bush brought the Republican Party to power in both chambers of Congress.

In a warlike mode, the deep philosophical thinking of an Ivy Leaguer could have become a handicap in communication with the people. So when someone on the street said that President Bush was hard-nosed and tough, they believed that apart from dealing with Saddam Hussein and Al-Qaeda, he was capable of attacking domestic issues as well, which they framed in broad terms as nothing but security—national, homeland, and economic security. He blamed the Democrats for holding him back from establishing an overarching, unified Department of Homeland Security "that will vastly improve our borders and our communities." Addressing a postvictory press conference in 2002, Bush said, "The election may be over but a terrorist threat is still real."[9]

Because of the midterm defeat, the Democrats withdrew their opposition to the homeland security bill. The creation of the new Department of Homeland Security involved the merger of twenty-two different departments and took political ingenuity and determination to bring about complete integration in a short time, in spite of doubts and apprehensions expressed by experts.[10] But the Bush administration could not wait that long for the new department to take shape when the immediate purpose was to use the collective strength of the new system—since the emergent system is greater than its subsumed parts—to do predictive intelligence so that terrorists could be liquidated before they attacked.

Clubbing an innovative system out of the existing ones is an arduous task, which was one of the reasons the Bush administration wanted a free hand in engineering the new antiterrorism homeland security department by suspending civil service rules, much to the chagrin of the Democrats and their biggest constituency, the labor unions. Whether hiring and firing employees based on performance rather than civil service rules would have helped the Bush administration to accomplish in a year what would have otherwise taken five years or more was open to speculation. The meta-agency, the Department of Homeland Security, created its own space-time dimension and worked at its own pace, as all new systems that arise out of mergers do. Not all mergers are successful, as can be seen in the private sector, but a warrior president who did not have the luxury of time on his side succeeded where others might have failed.

WAR GAMES AND DIPLOMACY

The United States was right in demanding that Iraq offer irrefutable evidence that preparations for weapons of mass destruction had been halted permanently and the infrastructure dismantled. As if in response, Iraqi general Amir al-Saadi said that the twelve-thousand-page report submitted to the UN Security Council was "entirely accurate, it's truthful."[11]

Producing such a voluminous report containing complex and confidential data about the process and technology of assembling nuclear weapons within the thirty-day deadline was itself a remarkable feat. In diplomatic appearance at least, Iraq was not in total defiance of the world community. General al-Saadi, a top adviser to Saddam Hussein, challenged anyone to prove that the report was not accurate. But then accuracy and truthfulness do not always create trust. British foreign secretary Jack Straw was quick to react and called the declaration "a pack of lies" without even making a preliminary examination of the document.[12] On the other hand, Mohamed ElBaradei, the UN's nuclear inspection team leader, sounded more sensible when he said, "We are going to begin immediately the painstaking and systematic process of checking the veracity of the document against the facts that our inspectors find on the ground."[13] It seemed that the diplomatic exchange would continue endlessly. "If they have anything to the contrary, let them come up with it to the Security Council," the Iraqi general said, implying that the United States and Britain should not take international law into their hands.

The due process, which is a revered legal tradition in America and wherever the rule of law prevails, should not have been ignored at the international arena, when the lives of millions were involved. To forestall any criticism, the general admitted that Iraq did have the means and the intentions, way back in 1991, at the time of the Gulf War, to produce nuclear weapons but gave up such plans. General al-Saadi said, "We have the complete documentation from design to all the other things. We haven't reached the final assembly of a bomb nor tested it."[14]

If the United States had a smoking gun, compelling evidence of the kind President John Kennedy had about the Soviet missiles based on the American U-2 spy plane that secretly photographed nuclear missile sites being built by the Soviet Union on the island of Cuba, the Bush administration should have given it to the Security Council before any action was contemplated against Iraq. In fact, for a number of years the Bush administration labored hard to build a case for invading Iraq.

In February 2003 Secretary of State Colin Powell gave an exuberant address to the Security Council during which he showed satellite photos in order to prove that Iraq had WMDs. The satellite evidence was spurious, but the United States had made up its mind to invade Iraq any which way.[15] On October 6, 2004, Charles Duelfer, the head of the Iraq Survey Group, informed the U.S. Senate Armed Services Committee that ISG had found no evidence whatsoever and, moreover, since 1991, when UN sanctions were imposed, Iraq had not produced any weapons of mass destruction.[16] UN Secretary-General Kofi Annan condemned the war as illegal because it was "not in conformity with the Security Council."[17] Ironically, it was only after the invasion that the Central Intelligence Agency, Defense Intelligence Agency, and other intelligence agencies distanced themselves from the evidence of Iraqi WMD or the regime's link with Al-Qaeda.

The U.S. news media had developed a submissive, collective mind-set and succumbed to the Bush administration propaganda machine. Subsequently, the collaborators, the Bush and Blair administrations, discovered other justifications for the war; for example, the Iraq government's gross human rights violations and the need to bring freedom and democracy to the Middle East.

The impression about the United States being an international bully rather than a gentle and benign superpower hardened in the minds of not only the Arabs and Muslims but also in other parts of the world where Americans have been generally perceived as friendly people. The whole idea of carrying out creative destruction through a fast-paced surgical operation to bring about regime change and rebuild a more acceptable and friendly Iraq was preposterous, dangerous, barbaric, and neoimperialistic.

Colin Powell, Secretary of State under the Bush administration, has not been able to clean himself of the blood on his hands and redeem himself of the factoids and half-truths he told an international audience listening to him at the UN Security Council, the fake account of WMD of which he was unaware but should have known, something expected of a most brilliant four-star military general. In spite of all the best-selling books, the poor general, like Lady Macbeth in *Macbeth* (Act 2, Scene 2) will be always saying:

> Will all great Neptune's ocean wash this blood
> Clean from my hand? No, this my hand will rather
> The multitudinous seas incarnadine,
> Making the green one red.

Not all the rag to riches to glory stories end in grace. The Iraq War eliminated a most cruel Arab dictator Saddam Hussein, but it also ruined the reputation of General Colin Powell, the African American who made it to the top but then stumbled and crashed. In search of glory, the general just could not see the truth.

AWESOME TEMPTATIONS OF NEOIMPERIALISM

Having won the war, the temptation for the United States to have some form of permanent presence in Iraq has been great because of the country's huge oil reserves, 150 billion barrels, one of the largest in the world. For Americans, some argue, it is crucial to break the Saudi stronghold on the oil supply because it creates a feeling of helpless dependency on a country that spewed up fifteen of the nineteen terrorists on September 11, 2001. With Russia, too, entering the oil market in a big way and post–Saddam Hussein Iraq opening up its oil fields to Western oil companies as well as China, some commentators thought it would be possible that the power of the Saudi-led OPEC cartel could be weakened if not totally broken.[18] The Arab Street couldn't have been more amused since the rest of the world helplessly depends upon the flow of oil to them. Besides, international carpetbaggers and born-again imperialists cannot give up on the Middle Eastern oil wealth.

British financial historian Niall Ferguson argued that the United States might not have much choice but to assume the imperialist mantle to maintain peace in the world. "True, the war against terrorism has a novel character—it is remote both geographically and technologically. It will nonetheless be Clausewitzian in principle, the wholehearted pursuit of a legitimate political objective by, regrettable but necessarily, violent means," Ferguson said.[19] Carl von Clausewitz (1780–1831) was a Prussian general, whose posthumously published treatise, *On War*, advocating a policy of total war, had a transformative influence on European military thinking, especially in Germany, with terrible consequences for Europe and rest of the world.[20]

Ferguson argued that the United States should have done what the British did during its Victorian heyday when Britannia ruled the waves unchallenged. In fact, Americans could do better, he argued, because while old Russia and Britain failed to achieve their goals in Afghanistan, the United States speedily defeated the Taliban and established its military and political presence in the region, as subsequently it tried to do in Iraq. In his continuous historical superspeculation and futurecasting, Ferguson has been warning the world of an impending great train wreck, if the United States of America did not mend its irresponsible fiscal behavior. Ferguson may be right, albeit for different reasons. Who will you call for help if America is down and out? There is no one else besides America that can mobilize its unfathomable energies—the strength that arises from the First Freedoms—to meet global challenges at a moment's notice. To let America decline and fall is immoral, inhuman, and unforgivable.

But the neoimperialist view of the world political situation is different in the sense that a most dominant political power, at present the sole superpower America, whose 2012 to 2013 GDP is more than the total GDP of the next two economic powers combined, must maintain world peace at all costs. "Since 9/11 there has been an unmistakably Clausewitzian flavor to US foreign policy," according to Ferguson, especially when you consider the Bush doctrine of pre-emption. "Liberals, most prominently will fret about the violation of national sovereignty, enshrined in the UN charter. But it worked for the British," he exulted unapologetically.[21]

As a historian Ferguson should know that it did not work always. Consider the 1956 French-British-Israeli attack on Egypt to wrest back the control of the Suez Canal, which ended in disaster.[22] Or the Vietnam War that has left deep scars on the American psyche. The Rumsfeld Doctrine—shock and awe, search and destroy the enemy by using smart digital, technology-based supercombat systems including drones, nevertheless, with limited reliance on small but dexterous ground forces—helped to win a quick victory, though without peace; it's doubt if it would work in more complex situations.[23]

Imperial Britain could ignore world opinion, which was not as well articulated as it is today in the age of global mass media and the Internet. Anglo-American neoimperialists wearing a shining armor of imperial history need to think again. "The existence of a military 'hyperpower' that really means business—that is able and willing to use its superior force—may in fact be better for world peace than any number of international treaties," persists Ferguson relentlessly in his historical infallibility.[24]

The doctrine is unworkable if we keep 9/11 as a focal point of modern times. The United States, for example, succeeded in buying support from the former Pakistani dictator General Pervez Musharraf and his ruling clique in fighting the Taliban and Al-Qaeda, but it has not been able to win over the Pakistani people, without which there can be no lasting peace and reconstruction in Afghanistan.[25] Then there is North Korea—China's pit bull terrier—whose nuclear defiance shouldn't be puzzling. Can the United States impose its will upon North Korea, China's partner in peace and harmony in Asia? Or Iran, a country that does not have a dictator like Saddam Hussein for America to demonize, but it is rather a nation that is ruled by a kind of Shia mullah-theocratic-democracy? Iran and North Korea are not the only two problems. China looms large in East Asia, where the neoimperialistic throwback-to-history doctrine would be disastrous. The patterns of history exist only in the minds of historians.

BEWARE THE POWER OF THE UNEXPECTED

War is brutal; we are reminded every time diplomacy fails or is not allowed to work. But its brutality becomes more palpable when someone you know personally is going to the front line and there is the chance that he may not come back. Thousands of families in the United States pray for their young men and women to come back safe. Wouldn't it be wonderful if somehow we could know how to get rid of the evil people amid us without hurting the innocent? Maybe future GPS-guided precision weapons of war, future sentient drones, will shock and awe, search and destroy only the evil ones without killing the innocent civilians. It may be possible in the future that Big Data research will help us to segregate people with evil intentions and apprehend terrorists before they bomb, for example, a most peaceful event like the Boston Marathon. The power of the unexpected is terrible.

On the first day of the invasion of Iraq in 2003, when skies over Baghdad lit up with fireworks of precision bombs, it looked as if we were playing some kind of video game with a remote. It was expected that the war would end quickly like the surgery done on a shocked-and-awed etherized patient, and peace would descend swiftly on the land once described as the cradle of civilization. Then came the news of some U.S. soldiers being shot in the head execution style and their gruesome bodies displayed as trophies on Al Jazeera and other Arab television networks. A nauseating feeling began to sink in. It might become a prolonged war, and the American public might not have a stomach for it.

Television and the Internet make war unbearable because they go on drumbeating the same visual message in a ceaseless stream. President Bush admonished the American people, “It may seem like a long time because of all the action on TV, but in terms of overall strategy, we’re just in the beginning of phases.”[26] Many Americans who supported the war thought it would be an unhindered race toward Baghdad with tanks and artillery protected by a massive air and missile umbrella. Even Wall Street cheered up lustily for a couple of days before it retreated grievously, as it realized with fear that the war might drag on and on.

Capitalism is timid, unless there is money to be made out of war. But Wall Street would surge again on the hope of cheap and abundant oil once the prospects improved. Whether the war is just or not, the central objective, regime change and disarmament, remained unchanged. As Vice President Dick Cheney told NBC’s Tim Russert, “We have to be prepared now to take the kind of bold action that’s being contemplated with respect to Iraq in order to ensure that we don’t get hit with a devastating attack when the terrorists’ organization gets married up with a rogue state that’s willing to provide it with the kinds of deadly capabilities that Saddam Hussein has developed and used over the years.”[27] That was the expectation, but it is the power of the unexpected for which it is difficult to prepare the people.

One of the most unexpected events of the war was the fratricidal attack when a U.S. Muslim soldier, Sergeant Hassan Akbar of the 101st Airborne Division, turned against his own brethren and threw grenades into their tents, killing two and injuring fourteen. What could have gone through the mind of Akbar in the heat and sand of the desert? If he thought the war was unjust, he wasn’t the only one. Michael Moore, the best-selling author of *Stupid White Men* and an Oscar-winning documentary filmmaker, shocked the Oscar night audience with his outburst: “We live in the time where we have fictitious election results that elect a fictitious president. We live in a time where we have a man sending us to war for fictitious reasons. Whether it’s the fiction of duct tape or fiction of orange alerts, we are against this war, Mr. Bush. Shame on you, Mr. Bush, shame on you. And any time you got the Pope and the Dixie Chicks against you, your time is up.”[28]

Can you imagine a man such as Michael Moore speaking so brazenly about the head of the government and living freely in a country like China or Saudi Arabia? It is only in the United States that a political structure of divided powers and fearless freedom is capable of setting in self-correcting processes to minimize authoritarian tendencies, which makes one hopeful that the U.S. imperial eagle will never victimize the world. Let’s bear in mind that the Pope, the Dixie Chicks, or Michael Moore can do no harm to any president. It is Wall Street and the economy, every president’s Achilles’ heel, which may be their undoing. There lies the power of the unexpected over which the U.S. president has no control.

BUILDING PILLARS OF FREEDOM ON SOFT POWER

In the prevailing state of chaos, violence, and daily killing, one wondered what the United States was trying to accomplish in Iraq and Afghanistan, apart from restoring law and order. Building roads, bridges, and telecommunications is necessary, no doubt, but the multibillion-dollar infrastructure, financed by the American taxpayer, could eventually become the paraphernalia of a repressive regime rather than a civic society, unless structures of freedom were also built. Saudi Arabia is an example of a country with the most modern infrastructure, gleaming highways, luxury hotels, and resorts, but it is a repressive society that uses its oil wealth for spreading its version of extremist Islam, Wahhabism, through madrassas, charities, and religious police.[29]

In order to build a modern civic society, it is necessary to erect a structure that stands on three immovable yet dynamic pillars: freedom of expression, freedom of and from religion, and most importantly, equal opportunity for women. Without women being free and equal to men in all respects, freedom cannot survive for long. Nor can family life be good without women being treated equally at home and in the workplace.

Without the structure of freedom, the three essential institutional bulwarks of human freedom, a country may become economically prosperous, may have open markets and free trade, but still could have closed minds fit for tyrants to rule. The magnitude of the problem that the United States has been facing for nation building is tremendous. It amounts to bringing about a massive cultural shift. For example, how would the United States persuade Afghans that freedom of thought includes the questioning of Islamic orthodoxy, abolition of laws relating to blasphemy, and giving women equal rights? Certainly it cannot be done overnight, but one might argue that the seeds must be planted now. The constitution of a country must be its instrument of freedom. So when a new constitution is framed, the most important thing is the inclusion and the supremacy of the fundamentals of human rights—free expression, freedom of worship, and equality of the sexes—along with other checks and balances that limit power and ensure accountability.

These aren't American values, but time-tested universal values that actually work to expand possibilities for human rights, economic growth, and self-renewal. For a country with religious and ethnic divisions, constitutionally guaranteed human rights along with divided powers is the only way to keep the country together. Power sharing is the best way to manage diversity, control chaos, and establish the rule of law to turn, for example, Afghanistan, Iraq, or newly liberated Arab Spring countries into open, modern societies. It is possible for a country to become economically and militarily strong nonetheless with its heart and mind still remaining barbaric.

To prevent Afghanistan from regressing to Taliban Islamic extremism, strong legal and political structures ensuring shared governance and equal opportunities within the framework of human rights must be carved in stone, into the constitution. It is this context that makes it so important to understand why the U.S.-led international military presence might become necessary for a number of years to oversee that the constitutional framework does not break down.

Whether other countries help or not, the United States must relentlessly pursue its goal of freedom and human rights as part of its international efforts in building civic societies everywhere. That's what gives the United States the moral authority to be accepted as a benevolent superpower. Even a poor country can afford to have fundamental rights. It is absurd to argue, as some Americans say in defense of China, that a certain level of economic prosperity is necessary before human rights can be made paramount. The rising tide of economic prosperity in China is not leading to a rising tide of human freedom. Consider why Hillary Clinton's book, *Living History*, was censored in Chinese translation.[30] How will China's authoritarian government accept freedom of thought and religion in Tibet or Xinjiang when it cannot trust its own people, the Mainland Chinese?

Why should we bother about human rights and freedom in China, the manufacturing powerhouse of the world? It is for the same reason we should be concerned about human rights in Saudi Arabia, Iraq, or Iran, countries with tremendous oil wealth, or Afghanistan with its abundant poppy crops. Sometimes suppression at home leads to terrorism abroad, if the outside world does not hear the cry for freedom. If Saudi Arabia were an open democratic society, 9/11 could not have happened. It's only through the slow and steady effect of soft power that the United States can open the doors of closed authoritarian societies. Soft power is not a miracle drug or a magic bullet, but it is irresistible and subversive.

HOW DOES SOFT POWER ARISE?

Soft power arises from the attraction of a nation's culture, technology, and values, the indescribable aura that surrounds it like a halo. American culture is a culture of openness, freedom of expression, and open roads that lead to the free marketplace of goods and ideas and innovation. Although the first impression most people around the world have of American culture is that it is the culture of Hollywood, pop music, and TV programs saturated with sex, violence and adventure, and that is not untrue. But essentially it is a culture of aspiration that holds the possibility of expanding human horizons.

According to a Gallup poll in 2012, of the 150 million adults worldwide who would like to migrate to the United States, twenty-two million were

Chinese, making America the most attractive destination not only for rest of the world but also for the Chinese who see their country becoming a dominant power in the next decade or so.[31] As *China Times* reported, "Some 46% of Chinese whose assets exceed 10 million yuan are considering a departure abroad, while 14% have already initiated additional steps in this direction, according to the survey conducted jointly by Hurun Report and the Bank of China."[32]

China has latched on to only a limited aspect of the U.S. culture, capitalism (state controlled) and free marketplace (export oriented), and it has become the fastest growing economy. If the United States and the free democratic world were to shut their doors on China, hypothetically speaking, the Chinese export-driven economy would shake up like a house of cards. Americans know that China cannot stop there. It will eventually open itself to other cultural influences including free expression and democracy, which is an example of how, by opening its markets to China, the United States has exercised its soft power and transformed a hostile nation that once called America a paper tiger to a neutral if not a friendly global neighbor.[33] Although pollsters tell Americans that they are the most disliked people in the world, it may not be the whole truth. More than 2.5 billion people living in India and China do not dislike Americans. They might disagree with some U.S. policies; nonetheless, they are fascinated with American society. Professor Joseph Nye of Harvard believes that a country can become attractive by co-opting people rather than bullying them. Just as trade with China and rising prosperity has co-opted the Chinese people and has given them new hopes and new dreams, a similar policy might transform Cuba, North Korea, and Iran, too.

In his article "The Misleading Metaphor of Decline," in the *Wall Street Journal*, Nye talked about engagement as the best way to exercise U.S. power. He suggested that international influence "comes from an effective aid and information program abroad. What is needed is increased investment in 'soft power,' the complex machinery of interdependence, rather than in 'hard power'—that is, expensive new weapons system." Again in another piece in the *International Herald Tribune*, Nye acknowledged that in order to fight terrorism both hard and soft power are necessary but "attraction is much cheaper than coercion, and an asset that needs to be nourished . . . From Hollywood to higher education, civil society does far more to present the USA to other people than the government does."[34] So does the boldness of an American president, who in the midst of daily terrorist bombing attacks flies to Baghdad to celebrate Thanksgiving with U.S. troops, or to Kabul, to observe the anniversary of the dare-devil killing of Osama bin Laden in his Pakistani hideout—the kinds of deeds that transform soft power to smart power.[35]

It is not only the glamour of New York City, Hollywood, or Silicon Valley that makes the United States the most appealing country today. Corporate America, in spite of its recent excesses and egregious conduct, also makes America attractive. When an American apparel maker opens a factory in Vietnam or Bangladesh, for example, it creates new hopes and dreams for the people. And that's how soft power is exercised, through the Walmart global-supply chain as well as through Apple, Google, Facebook, or Twitter global-supply chains of ideas. But what could be a greater expression of soft power than to demonstrate to the world that the dream of the aspirational society anchored in the First Amendment is for everyone who calls himself an American. And to pursue that dream of the aspirational society, immigrants and their children have as great a future as anyone else in the United States.

THE FIRST AMENDMENT AND THE ASPIRATIONAL SOCIETY

Barack Obama's rise in the U.S. political firmament has been phenomenal. As he was seeking the presidency in 2008, many commentators saw him metamorphosing the American society, reminiscent of John F. Kennedy: "My fellow citizens of the world: ask not what America will do for you, but what together we can do for the freedom of man."[36]

Obama's soaring eloquence about hope and change sometimes echoed Reverend Martin Luther King Jr., who audaciously hoped that one day the United States will become a just society. But apart from idealism, there are other systemic forces at work. An open democratic society like the United States that thrives on innovation and marketplace competitiveness needs the best to rise to the top to manage its affairs. Race and gender are handicaps to the merit-based aspirational society, but gradually they have been becoming less important. The United States is the place to dream impossible dreams and make them real, but it is the dynamic system built on the First Amendment that encourages people to imagine the unimaginable.

Everyone talks about the American dream, but what is it? The American dream is a heightened state of aspiration that drives a person to break barriers and achieve their goals, regardless of the background the person comes from. The emphasis is upon hard work, ingenuity, self-education, and courage to take risk. It is challenge of the free marketplace: compete and create space for yourself. You could see it in the life of a man like Andy Grove who escaped the Nazis and the communists and came to the United States as a refugee.[37] He built a microchip empire, Intel, the technology that underpins the information superhighway and serves America's global interests. But he did not do it through the benefits of affirmative action and racial quotas. Nor does Intel, of which he was the CEO, hire people based on quotas, racial preferences, compensatory guilt, or the need for diversity. The aspirational society thrives on competitiveness, not on affirmative action.

A JUST AND MERITORIOUS SOCIETY

But don't affirmative action preferences create a just society? It has been argued that the rise of Colin Powell, who reached the top echelon in the U.S. military with an extraordinary record of achievements and served the Bush administration as Secretary of State in a very difficult era of American diplomacy, wouldn't have been possible without affirmative action. But it is also true that though affirmative action might have opened the door for General Powell, it did not push him to the top and make him one of the most esteemed American soldier-diplomats. In his book, *My American Journey*, he wrote, "Equal rights and equal opportunity . . . do not mean preferential treatment. Preferences, no matter how well intended, ultimately breed resentment among the non-preferred. And preferential treatment demeans the achievements that minority Americans win by their own efforts."[38] Consider, for example, the achievements of stage and screen actor and author Bill Cosby, pop culture's global icon Michael Jackson, and basketball's most glorious athlete Michael Jordan, people who rose to fame and fortune on their guts and talents in an open and competitive marketplace.[39] Such remarkable achievements would be diminished if they were associated with racial preferences and quotas.

What Americans ask for is a level playing field to build their dreams on. As General Powell said, "If affirmative action means programs that provide equal opportunities, then I am all for it. If it leads to preferential treatment or help those who no longer need help, I am opposed. I benefited from equal opportunity and affirmative action in the Army, but I was not shown preference."[40] General Powell rose to the top on the strength of his character and brainpower in an institution—the U.S. military—that thrives on these qualities. But what about those who live in ghettos, inner cities, and in rural poverty? "If a history of discrimination has made it difficult for certain Americans to meet standards, it is only fair to provide temporary means to help them catch up and compete on equal terms. Affirmative in the best sense promotes equal consideration, not reverse discrimination," wrote General Powell.[41]

In spite of the spectacular rise of many African Americans, Latinos, and Asian Americans, race and religion do matter in America. Prejudice is widely prevalent, and sometimes it is the color of your skin, not the content of your character or merit, that determines where you live and work, how a policeman treats you in the middle of night when he sees you at a street corner, or whether, when you go to the airport, you are "randomly selected" for special inspection because you look like a Muslim or have a Muslim name.[42] But these are the imperfections of a dynamic aspirational society.

Let's keep in mind that even decades of affirmative action policies, which in reality amount to creating preferential treatment for minorities, have, however, not created a perfect color-race-gender-blind society. For a long time the dominant sentiment in the country has been to limit racial preferences because they create reverse discrimination. But diversity of race, religion, and gender and sexual preferences is socially desirable because it breeds new ideas that enrich society, and, moreover, it encourages tolerance and acceptance of America that is increasingly becoming multicultural and multiracial, as the 2012 presidential election amply demonstrated. The challenge is to use limited, just good enough, affirmative action to give some deserving people a bit of a head start without creating a sense of entitlements. The rise of former Secretary of State Condoleezza Rice, UN ambassador Susan Rice, and many others, on their own merits, will eventually make affirmative action less necessary.

But look at the millions of illegal immigrants from Mexico and elsewhere! They do not ask for any affirmative action or preferential treatment. They hope and believe in the essential goodness of America embodied in the First Amendment. All they ask is a chance to work, build a good life for their families, and legitimatize their presence as citizens. And in the process they enrich America's cultural diversity, enhance America's brainpower, and make the United States a most competitive, attractive, and desirable place on earth. That is what Barack Hussein Obama has done.

THE FLIGHT OF THE BLACK SWAN TO THE WHITE HOUSE

The whole world waited for president-elect Barack Obama, one of the most inspiring orators of modern times, to hear what he would say to the American people, the nation in deep economic distress, as he entered the White House on January 20, 2009. Perhaps no one would have been more gracious than President George W. Bush, who said, "It will be a stirring sight to watch President Obama, his wife, Michelle, and their beautiful girls step through the doors of the White House."[43]

I imagined President Obama hugging and folding his wife and children in his huge embrace with a broad smile, waving to the milling and swaying crowd, as if to say, "Thanks for the Goodness of America, the First Freedoms, the Aspirational Society, we have arrived at last, we have arrived at last." A generation ago, it would have been unthinkable that the American-born son of an African immigrant married to a white woman would have the courage, "the audacity of hope," to aspire to sit in the Oval Office. By winning the presidency in 2008, President Obama elevated the White House, uplifted the American people, and awakened the world to the limitless possibilities of the aspirational society. Many people hoped that he would transform the United States into a moral force, once again a great power for the global world, but that aspiration still remains alive.

One can only do superspeculation about the journey: from the White House the Obama family would have surveyed the black-and-white past, darkness to light, from the coming of Kunta Kinte (Arthur Haley's fictional-historical character from *Roots*) to America in a slave ship to work in the cotton fields of the South to his liberation after the Civil War, which soon mutated into the utter degradation of segregation through the Jim Crow laws; his slow march through judicial and legislative processes such as the 1954 Supreme Court decision *Brown v. Board of Education* declaring segregation as inherently wrong, and the 1964 Civil Rights Act outlawing racial segregation; and his final assertion and ascension as a human being and the aspiration about a new America where one's character and merit, as the nonviolent civil rights movement leader Martin Luther King Jr. prophesied, will matter more than the color of one's skin.[44]

In a manner of speaking, the history of America is the history of the black man rising and finally being accepted as a fellow American by the white man. Now that Kunta Kinte has entered the White House, this is the end of history, not the kind of end that Francis Fukuyama speculated about. It's the beginning of new possibilities at home for regenerating America through innovation, invention, and entrepreneurship, of establishing a new system of universal health care and financial security, of redemocratizing and rededicating the American dream that there is always a second chance and failure is no stigma, and yes, you can do it.

In the 2008 presidential election, Barack Obama built a remarkable grassroots, Internet-based, metaparty political machine to win the hearts and minds of the young and the old, of the entire racial and ideological spectrum, and got almost a mandate to bring about transformation in American society. But as the ravaging global financial crisis showed, Americans are so densely interconnected with the rest of the world that no meaningful change can occur unless the rest of the world is willing to cooperate, a lesson the United States should have learned after the 9/11 terrorists attacks.

The United States cannot go on managing global affairs alone, not even in its attempt to set its own house in order, without help from others. The United States has to learn how to deal adroitly with the badlanders of Pakistan, the Mullahs of Iran, the reckless adventurers of North Korea, and the amoral moneylenders and peaceful risers of China. So the question at the time of his inauguration was how Obama would transform his campaign slogan, "Yes, We Can," into ground reality.

When Obama, surrounded by some of the best and the brightest economists and financial experts, gave his first postelection press conference on

November 8, 2008, he was cool and measured in his remarks about the state of the economy and sounded much less exuberant about the prospects of turning things around than when he was stumping for the election. "I do not underestimate the enormity of the task that lies ahead," he said. "Some of the choices that we make are going to be difficult. And I have said before and I will repeat again: It is not going to be quick and it is not going to be easy for us to dig ourselves out of the hole that we are in."[45]

The most sobering lesson an American president has to learn is that he cannot do much about global markets. Being the leader of the free world is a most dangerous delusion. Whether there is an opportunity for finding a new leadership role for the United States in an interdependent world of rapidly emerging centers of economic power—China, the European Union, and Russia—only time will tell. But when a reporter asked how Obama would respond to Iran president Mahmoud Ahmadinejad's letter of congratulations, he said with a determined finality that Iran's nuclear ambitions were absolutely "unacceptable."[46] I hope that one day, sooner than later in his second term in the White House, President Obama will have the courage to say: The decline and fall of the United States of America is utterly "unacceptable."

In January 2009, Kunta Kinte had a royal feast to savor in the White House: global financial meltdown; the continuation of Israel-Palestine decades-long deadly dance; Iran's "peaceful" nuclear ambitions; Afghanistan and Pakistan implosion into failed states; and the dragon on the rise, fuming all over the South China Sea claiming sea beds and skies as its exclusive zones of national interest. Soon the reality set in for President Obama.

THE ASPIRATIONAL SOCIETY MUST NOT SHRUG

The great hope and optimism that Barack Obama, the 2008 presidential candidate, raised in the hearts and minds of his ardent worldwide supporters, who saw in him a liberator made like Abraham Lincoln, began to fade fast soon after he entered the White House, though the Cheshire-cat-like smile never left him.[47] Barack Obama turned out to be more like Cardinal Francesco Barberini in Bertolt Brecht's play, *Life of Galileo.* Cardinal Barberini, unlike the Catholic Church establishment, promised not to ignore science, not to "set myself up against the multiplication tables" and "the physical facts."[48] And when Cardinal Francesco Barberini is elected to become the pope, he raises great hopes in Galileo and the scientific community that the era of irrational fear is coming to an end and the boundaries of permissible thought would expand in Europe. But as Cardinal Barberini gracefully assumes the heavenly ordained vestments of the pope and gets ready to make a public appearance, he acquiesces in the demands of the Inquisitor that Galileo must face trial for his perfidy: the denial of the Church-established truth.

Pope Barberini embraces the Church fantasy of the earth-centric heavenly order as Kunta Kinte, now President Barack Obama, has embraced President Bush's rhetorical fantasy of saving the Homeland from the circle of evildoers. Barberini the pope and Obama the president were both swallowed by the system that they wanted to control.

There is hardly any difference between President George W. Bush's determination to destroy the "the axis of evil" and what President Obama said during his Nobel Peace Prize acceptance speech: "Evil does exist in the world. A nonviolent movement could not have halted Hitler's armies. Negotiations cannot convince al-Qaeda's leaders to lay down their arms. To say that force is sometimes necessary is not a call to cynicism—it is a recognition of history, the imperfections of man and the limits of reason."[49] President Bush himself couldn't have phrased it better. Just as Galileo and the scientific community were let down by Cardinal Barberini when he became the pope, many Americans and people abroad who had ardently supported Barack Obama felt squashed by him as soon as he entered the White House and became the embodiment of the American political system and the dominant ideology that the United States must fight evil abroad and America must triumph. The system triumphs again and again!

EMPATHY AND IMPERIALISM

I looked around for help and found deep understanding and profound insights in the work of a humanistic scholar and literary critic, Donald E. Pease of Dartmouth, whose views are radically different from the neoimperialistic historian Niall Ferguson's views as to what America should do to the world as a global power. Great historical crises have to be narrated in a mythological language to make them understandable and acceptable to people in a profoundly significant and moral way. The persuasive power of the overarching and governing myths, the metanarratives, like the Virgin Land, Ground Zero, and Homeland, comes from their ability "to produce and conserve national traditions" and therefore "to function as the unacknowledged legislators that regulate people's thought and behavior" expounds Pease, in his thought-provoking book *The New American Exceptionalism.*[50] The myths of the Virgin Land, the Wild West, the New Frontier, and American Exceptionalism resound and reverberate throughout Hollywood and popular culture, as well as American history, giving rationale and moral justification to the domestication and pacification of Native Americans, inoffensively termed as independent sovereign nations, by putting them into the wilderness and treachery of reservations. "The Virgin Land" stands for an America that is an "open national landscape that fostered the construction and realization of self-reliant individualists," a fundamental bonding fantasy that has amalgamated the people with their land.[51] Fantasy is an obsessive and recurrent desire that seeks self-actualization, as happened in Nazi Germany and is perhaps happening in China today.

But in the long run no all-encompassing fantasy can turn a people into an animal farm, a buffalo herd forever, especially the American people drunk deep on the First Freedoms and given to iconoclasm and idol breaking. Divisiveness, skepticism, distrust, and cynicism are natural to American character, and to a great extent this trait motivates them to search, seek, and innovate. Nonetheless, Pease's analysis of what has happened to post–Cold War and 9/11 America is fascinating. The abrupt end to the Cold War and the consequent disappearance of the Soviet Union from American consciousness exposed a chilling geopolitical fault line in American society, which, though not visible during the George H. W. Bush administration due to Operation Desert Storm, ultimately found strong expression in two opposing visions of the United States: Newt Gingrich's "Contract with America" enfolding the red states and Bill Clinton's multiracial and multicultural "New Covenant" covering the blue states. The government's mishandling of the fundamentalist Branch Davidian church at Waco in 1993 and two years later the Oklahoma City bombing showed a nation violently divided against itself. "Both of these events were embedded in the apocalyptic dimension of state fantasy of American Exceptionalism, but after the cold war American Exceptionalism ceased representing all the people," argues Pease.

The shocking 9/11 terrorist attacks closed the fault line in the nation's geopolitical psyche and provided the state with an opportunity to mythologize the World Trade Center site as "Ground Zero," perhaps a foretaste of Hiroshima, what the "Other" might have done to the United States of America if it had weapons of mass destruction (WMD). "The Shock and Awe" was a brutal response to the attack on Americans' self-image of being an exceptional people, people of the Virgin Land.[52] The horrific event returned to the American people their "collectively shared state fantasy" that united them once again, as they had been during the Soviet era Cold War.[53]

The Bush-generated state fantasy of protecting the Homeland by waging a global war on terrorism muffled the Clinton-Gingrich divisive politics and temporarily reunited America. But at the same time, it transformed the nation from citizenry into a domestic protectorate, similar to what Supreme Court justice John Marshall in 1830 had done to Native Americans' homeland territories. "The official narrative accompanying Ground Zero has linked the people traumatized by the events that took place on 9/11 with a Homeland Security State that emerged with the loss of the belief in the inviolability of the Virgin Land," reasons Pease, examining America through a sociopsychoanalytic and cultural criticism perspective, very different from Ferguson's neoimperialistic Clausewitzian vision of America.[54] To fight the enemy, the violator of the myth of Virgin Land and American Exceptionalism, the Bush administration imposed emergency rules by enacting the USA PATRIOT Act, which abridged freedom and shortchanged the rule of law—the very law to protect which President Bush had determined would best fight the enemy.

Legally and politically reframing Guantanamo Bay as a foreign country empowered the Bush administration to place it beyond the jurisdiction of U.S. courts, thereby turning the inmates into "unlawful enemy combatants," outside the protection of the Geneva Conventions. By generating such "master fictions," the Bush administration succeeded in concocting public consent, thus making the American people implicitly complicit in the state's violation of the rule of law. By misappropriating "We the people" into the Homeland Security State, it diminished the American people of their ability to think independently as citizens, postulates Pease. But fortunately this collective state of mind did not last long.

Propaganda

How the Bush administration's contrived propaganda succeeded in bamboozling the tripart structure of coequals in power and the supposedly fiercely independent news media, including the *New York Times* and the *Washington Post*, remains to be explained. But the news media somehow or the other did wake up from their doped state and asserted their freedom on April 28, 2004, when CBS's *60 Minutes* broadcast the horrific pictures of Abu Ghraib prisoners. The images jolted the American people from their numbed state. It seemed that American troops fighting in Iraq were repeating the acts of violence that had been committed against the slaves. The brutal, ugly past Americans had been trying to disown had suddenly burst through the depth of the subconscious topography. The racial hatred and cruelty had not totally disappeared, it seemed. "Man on the box" looked to be an object of sacrifice and also a slave being lynched. "The images mobilized opposition to the occupation in the name of Abu Ghraib prisoners who had been victimized by this oppression," writes Pease.

The discovery of Abu Ghraib atrocities was a turning point in the American mode of self-consciousness, their sensibility about their own sense of exceptionalism. Exceptional people do not indulge in such extremely degrading cruelties even against their worst enemies. Unfortunately, it seemed that the war against terrorism was being fought through culturally degrading images. Although "the pictures of these repugnant forms of violence" created a deep sense of shame and sorrow and embarrassment, the support for war would continue, "as long as U.S. publics remain captivated by the spectacles of violence the state had erected at the site of Ground Zero."[55] Pease admonishes: "If U.S. enemies were so 'barbaric' that they would not surrender even in the face of these acts of degradation, U.S. troops were in it for a long haul."[56] How true even today!

So strong was the hold of the Homeland–the Virgin Land on the people's imagination that they continued supporting the war until they heard the heart-wrenching cry of Cindy Sheehan, the mother who lost her son in the war. Sheehan repudiated the Iraq War that "redefined maternity as sacrifice for the cause of the nation."[57] Support for the war had been lessening. How much of disenchantment about the war was due to Cindy Sheehan, whom Pease compares with Greek dramatist Sophocles' Antigone's act of defiance and civil disobedience against the authority of the state that had refused her rebellious brother a proper burial, it is rather hard to say. But along with Abu Ghraib and the nonexistence of weapons of mass destruction, the Bush-administration-created rhetorical fantasy had begun to disintegrate.[58]

Donald Pease and Niall Ferguson look at America in different but fascinating ways. Despite 9/11 and a decade of terrorism, war, and bloodshed everywhere, the American sense of exceptionalism born of the First Freedoms remains unscathed. President Barack Obama, like other aspirants to the White House, has unabashedly and enthusiastically embraced the myth, the fantasy, and the metanarrative of American Exceptionalism—"*the American dream, the perfectible Union, and the land of promise.*"

The land of promise, the United States of America, like Atlas, must accept cheerfully the moral responsibility of "having the weight of the world on his shoulders."[59] If America shrinks, who will nurture the Arab Spring? Atlas must not vacillate. The aspirant society must keep its mind and doors open.

Conclusion: Free Speech as Energy

PUSHING THE BOUNDARIES OF PERMISSIBLE THOUGHT

The great archeologist and historian Ian Morris argues in *Why the West Rules—For Now: The Patterns of History, and What They Reveal About the Future* that Western dominance, keeping in mind more than ten thousand years of human history, has been only for a brief period of the last few hundred years, especially since the beginning of the Industrial Revolution in 1750 when steam and coal energized the West to colonize the rest of the world. The rise of the West has been not due to some intrinsic racial or cultural superiority but because of extraneous geographical factors and how human beings respond to them. The pattern of history is immutable, and geography matters more than human agents in determining the pattern of history, he argues, because "the force driving the rise of the East is exactly the same as the force that drove the earlier rise of the West: the interaction of geography with economics and technology."[1]

But unfortunately Morris does not consider it significant that the rise of the West ran parallel with the expansion of the boundaries of the permissible thought and free speech made possible by the development of the Gutenberg printing press, the Renaissance, the Reformation, the Enlightenment, and the Emancipation Proclamation. Culture does matter. For example, one cannot stop wondering as to why the moveable type printing press, which was first developed in China in 1040 (although the *Diamond Sutra* was published in AD 868), had so little effect upon China and the East than it had in Europe where the first moveable printing press was developed in 1436 (the Gutenberg Bible was published in 1455). The printing press liberated the European mind from the doctrine of infallibility and medieval papal authoritarianism, and it enabled Europeans to rediscover and reclaim the Greco-Roman spirit of rational inquiry through free speech (*Nullius in verba*).[2] When you take no one's word for it, for anything, an iconoclastic attitude born of the culture of rational inquiry and the spirit of challenge and response, you are free to break the pattern of history, and history ceases to be a "self-organizing system of autonomous agents" or material forces. Left to natural forces, social systems, like species, could go on living forever. Jellyfish have lived for five hundred million years without blood, bones, or brain. Native American societies would have continued their existence more or less undisturbed had the Western exploration, conquest, and colonization not intervened. India's caste system is still alive (albeit it is dissipating in the wake of rapid urbanization). But slavery in the United States was brought to an end through free speech, the rhetoric of freedom and equality that led to the Civil War.

Some time in the history of humankind development, free and fearless speech became a determining force of evolution by enabling humans to explore beyond "the adjacent possible" and tinker and innovate new futures, as one sees today in the development of thought-controlled robotic prostheses. Scott Pelley of CBS *60 Minutes* reported an interesting case of a quadriplegic woman, Jan, who suffered from a genetic disease, spinocerebellar degeneration, that had paralyzed her from the neck down. Under the auspices of a Defense Advanced Research Projects Agency (DARPA) project called Revolutionizing Prosthetics, she volunteered to have sensors implanted in her brain, which were connected to a robotic arm. With the power of her thoughts only, Jan was able to move the robotic arm to pick up a cup of coffee and later shake hands with Pelley. Dr. Geoffrey Ling, a neurologist and retired Army colonel, who directed the $150 million project with a mission to create artificial limbs to help disabled veterans to live a normal life, expressed the hope that one day it would be possible to restore sight and hearing to humans. As he told Pelley, "We will have now created the possibility of restoration and we're not going to stop at just arms and hands. I dream that we'll be able to take this into all sorts of patients . . . with stroke . . . cerebral palsy . . . the elderly."[3]

This is an example of how fearless thinking born of free speech pushes the boundaries of permissible thoughts for which one cannot find patterns in history. Nor are the boundaries of permissible thoughts determined by geography, economics, or technology. As Stuart Kauffman says, "On the complexity scale of molecules, morphologies, species, Chevrolets, and legal systems, the universe is overwhelmingly non-repeating, or 'nonergodic.'"[4] Legal systems, Linux, iPhones, and NASA's Rover (*Spirit* and *Opportunity*), in the ultimate analysis, have their origins in a culture of free speech and free inquiry, in the spirit of Spinoza, Locke, and Voltaire. Interaction of geography, economics, and technology without free speech can lead only to one-party authoritarianism of the Soviet Union, Nazi Germany, and modern China. But China might have reached a turning point. As MIT professor

Yasheng Huang admonishes that China must "democratize or die," though some party loyalists believe that China has reached a postdemocratic future where democracy is unnecessary.[5] But to become a country as desirable to live in as the United States, or to excel in all the multifarious facets of its civilization, China will have to assimilate the essence of the First Amendment as its core value, without which China's growing rich will head for the exit to the West. In 2011 the illegal outflow from China, according to reports, was $472 billion, and 60 percent super-rich Chinese, those with assets of more than a million dollars, were considering emigrating. Chinese birth tourism to the United States and Canada has been rising.[6]

Immigration, and the consequent brain flow, is the best measure of the desirability and likability of a country. It is the United States, along with Canada, Europe, and Australia, that attracts the most people from all over the world. Martin Wolf argues in *Why Globalization Works*, "The liberating technological changes . . . reflected a new way of organizing the economic activities of society as a whole—a sophisticated market economy with secure protection of property rights. But freedom to seek one's own way in life, outside the boundaries of caste, class, community or, more recently of gender, cannot be restricted to economic activities alone. The culture of a liberal society is, for this reason, inimical to established hierarchies of power or opinion. It is no accident that commercial societies came to consider freedom of thought and expression of great value . . . The combination of practicality, rationalism and freedom of inquiry became the basis for the West's greatest achievement—modern science. It is again, no accident that science reached its greatest flowering in a commercial West."[7] Free and fearless speech is an indispensable source of energy for the development of science, technology, commerce, and most of all civic society, because it expands the frontiers of permissible thought.

Dilemma

The First Amendment, one of the greatest human utterances ever uttered, is a great disrupter, and it breaks the deterministic pattern of history. Free speech is an unfathomable source of vitality and is freely available; but the challenge is to harness it for constructive purposes so that the aspirational society does not reorder itself into an acquisitive or destructive society. Emergence of gangs and boardwalk empires of the early twentieth century in the United States, or the rise of the Third Reich and Khmer Rouge, or the 2008 Wall Street financial collapse, are examples of societies pushing their way into new orders, Kauffman's concept of "the adjacent possible," but not all emergent orders and organizations are equally desirable.[8] The corrective can come only from access to information, fearless speech, and robust news media, and most of all the media's capacity and the ability to know. Access to information without the corresponding abilities to understand it is the greatest threat to the aspirational society of today, a society that depends upon Big Data, whether it is Wall Street, global financial systems, health care systems, or terrorism. The future of liberal societies and their desirability may depend upon how free speech penetrates and exploits the ever-growing immensity of Big Data.

In the latter part of the twentieth century, the scope of the First Amendment was extended through the Sunshine Act and the Freedom of Information Act to give the American people access to government agency meetings and information otherwise deemed as classified. The right to know and access information, however, does not apply to private organizations, which compels journalists to be creative in developing confidential sources or to gather information through public sources. Most importantly, the right of access to information does not guarantee the capacity or ability to know, especially in the digital age of Big Data when a journalist is confronted with Wall Street financial complexities.

Financial journalists do what other journalists do—they report and analyze on the basis of information they gather or are given by their sources. But financial instruments and new innovative products have become so complex that very few journalists and experts understand them.[9] By and large, the brightest journalists are not attracted by financial journalism, and consequently Wall Street and corporate America go underreported or poorly reported. After Wall Street was hit with a financial tsunami in 2008, some financial writers such as Michael Lewis came up with fascinating 20/20 hindsight, fictionlike, character-driven stories as to what had happened to Wall Street. The impression one gets from these narrative accounts is that Wall Street is either full of rogues or incompetent people and one does not know who is who and therefore it is difficult to comprehend and predict when the next crisis will hit. If American journalists pay as much attention to Wall Street and global financial institutions as they pay to Hollywood gossip and sex scandals of politicians and sports celebrities, perhaps it would be possible to foresee and prevent the next crisis.

Wall Street, which has become a symbol of American economic power and a metonym for global financial markets, remains more or less outside of the U.S. government's control. While the tripartite system of a government of coequals in powers, the executive, the legislative, and the judicial, has maintained a system of checks and balances to minimize the abuse of power, Wall Street, on the other hand, is its own self-governing universe, an exemplar of interdependent coarising of an autonomous power structure, without any similar corrective abuse-of-power controlling system. Sometimes it seems that Wall Street has nothing to do with the United States, though it contributes substantially to New York City in terms of jobs and annual revenues. It attracts some of the best global brains whose actions, including financial innovations, are beyond the comprehension of most media people.

It is on Wall Street where the First Amendment, freedom of speech and of the press, does not always work. Protest movements such as Occupy Wall Street will never work. How can the physical occupation of Wall Street affect trillions of dollars of electronic capital flow through the system, which is global in nature? Wall Street's opacity is the greatest danger to the United States and the rest of the world. The Securities & Exchange Commission and other regulatory bodies are not enough. Wall Street desperately needs WikiLeaks-like whistleblowers that keep its internals open and transparent so that misdeeds, acts of commission and omission, are apprehended before they happen. Wall Street needs an early awareness system (EAS), not merely an early warning system (EWS), a system that works at the aspirational stage, the way the U.S. Homeland Security works to prevent terrorism at the aspirational stage. EAS deals with probabilities, not with actualities that are already taking place. The LIBOR (the London Interbank Offered Rate) interest rate rigging scandal that involved the Royal Bank of Scotland, Barclays, UBS, and other global banks, for example, could have been prevented if the equivalent of the Homeland Security early awareness system, an organization like WikiLeaks, or investigative financial media had been keeping a close scrutiny.[10] Unless free speech penetrates deeply into Wall Street, we will be moving from one bailout crisis to another, wondering what to do with these too-big-to-fail financial institutions that have become so indispensable to the global economic system.[11]

DEPENDENT ORIGINATION OF THE ASPIRATIONAL SOCIETY

Imagine a society that has all the trappings of a free democratic society, including free speech and freedom of the press, but the government controls the commanding heights of the economy and the marketplace. In the course of time the marketplace will become stagnant. The spirit of challenge and response can exist only if the marketplace is open and competitive. Such a society creates dependent classes. In such a society there is only one-way brain flow, only outward as it happened in India until the economy opened up in the early 1990s. It is possible for a closed or an authoritarian society to pump huge amount of funds and manpower into certain projects and achieve quick returns; for example, North Korea, Iran, and Pakistan, poor countries that have made major advances in missile and nuclear technologies. Such societies do not have a wide base of innovation, which can come only from fearless freedoms of the First Amendment. Religious orthodoxy and ideological doctrines, prisons of blind faith, prevent people from exploring the adjacent possible, break mental barriers, and innovate and aspire to the impossible. A fearful society, however nationalistic, cannot be sustainably innovative.

China's phenomenal economic growth has been the result of massive flows of capital and technology from countries where free speech is the norm—Taiwan, Japan, South Korea, and of course, from the United States—especially after the normalization of relations was effected by the Nixon-Kissinger visit in 1972. Hong Kong, with its highly developed financial market, English language and university education system, the rule of law, civic institutions, and cultural openness, was the greatest gift a colonial power gave to any country. Hong Kong, Japan, South Korea, Taiwan, and the United States made possible the spectacular growth of China. Imagine if China's eastern neighbors were countries such as Pakistan, Sri Lanka, Bangladesh, and Afghanistan! Or if Nixon-Kissinger had not opened up the United States to China! Capital and technology flow into China would not have taken place at the same rapid and sustained pace.

On the other hand, the aspirational society is a consequence of interdependent origination, coarising from the First Amendment, the culture of innovation, and free marketplace that encourages risk-taking behavior and attracts outside brainpower.[12] The aspirational society is a negation of historical patterns because it creates spaces for anyone with an entrepreneurial spirit. But entrepreneurs need the free marketplace, which cannot exist without free speech, the free flow of information, and the ability and capacity to know. Why does the free marketplace need free speech? This is so because without the right to know and the capacity to know, Wall Street and corporate America become a closed system, corruption possibilities increase, rogue traders arise, and the system crashes. Wall Street begins to look like Martin Scorsese's *Boardwalk Empire*, certainly a place with some order, however undesirable. Wall Street can become closed in two ways. One way of shuttering it is when access to vital information is denied. The other way Wall Street can become closed is when the system becomes so complex that no one or very few people understand it.

Free speech and free press cannot exist without a free marketplace. Government or philanthropy-supported expression cannot be totally free because it will not tolerate risky behavior. Brain flow can occur either through immigration or by building a global intellectual grid, but this can happen only if the marketplace is open and competitive. Free speech, free marketplace, and brain inflow create the culture of innovation, which feeds into the loop of interdependent coarising of the aspirational society.

The aspirational society is a dynamic process of interdependent origination. It arises out of four mutually dependent and challenging dynamic systems, including: (1) the First Amendment that creates spaces for innovative behavior by freeing us from irrational fears (whether it is stem cell research for life enhancement or redefining the idea of marriage by applying the

Fourteenth Amendment to gay marriage rights, for example); (2) an open marketplace predicated upon competition and subject to close scrutiny by the instrumentalities of the First Amendment; (3) the culture of innovation created by fearless freedom, iconoclastic attitude, and irreverent humor, and the opportunities of the free marketplace; (4) brain inflow through immigration and global networking and collaboration.

The aspirational society is a most desirable society because it empowers us "to hope till Hope creates / From its own wreck the thing it contemplates."[13]

Notes

INTRODUCTION: THE MAKING OF THE ASPIRATIONAL SOCIETY

1. Michael Greenstone and Adam Looney, "Ten Economic Facts About Immigration," Brookings Institution, The Hamilton Project, September 2010, http://www.brookings.edu/~/media/research/files/reports/2010/9/immigration%20greenstone%20looney/09_immigration.pdf.

2. Jack F. Matlock Jr., *Autopsy on an Empire: The American Ambassador's Account of the Collapse of the Soviet Union* (New York: Random House, 1995).

3. Jean Medawar and David Pyke, *Hitler's Gift: The True Story of the Scientists Expelled by the Nazi Regime* (New York: Arcade Publishing, 2001).

4. The Colbert Report, October 17, 2005, "The Word—Truthiness," http://www.colbertnation.com/the-colbert-report-videos/24039/october-17-2005/the-word---truthiness.

5. *King Lear*, Act V, Scene II, "Ripeness is all," http://www.bartleby.com/70/4352.html.

6. Vijay Govindarajan and Chris Trimble, *Reverse Innovation: Create Far from Home, Win Everywhere* (Cambridge, MA: Harvard Business Review Press, 2012).

7. Clayton M. Christensen, *The Innovator's Dilemma: When New Technologies Cause Great Firms to Fail* (Cambridge, MA: Harvard Business Review Press, 1997).

8. Steven Johnson, *Where Good Ideas Come From: The Natural History of Innovation* (New York: Riverhead Hardcover, 2010).

9. John Kao, *Innovation Nation: How America Is Losing Its Innovation Edge, Why It Matters, and What We Can Do to Get It Back* (New York: Free Press, 2007).

10. N. D. Batra, "Some Creative Messiness, Please," *The Times of India*, March 19, 2010, http://articles.timesofindia.indiatimes.com/2010-03-19/edit-page/28134589_1_foreign-workers-singaporeans-guest-workers.

11. Niall Ferguson, *Civilization: The West and the Rest* (New York: Allen Lane, 2011).

12. Mario Coccia, "Democratization Is the Determinant of Technological Change," Consiglio Nazionale delle Ricerche (Italia) and Max Planck Institute of Economics (Germania) CERIS-CNR, http://www2.ceris.cnr.it/ceris/workingpaper/abstract/wp2008-06.htm.

13. Lee C. Bollinger, *Uninhibited, Robust, and Wide-Open: A Free Press for a New Century* (New York: Oxford University Press, 2010), 5.

14. Zhiwu Chen, "Free Press Could Help China's Economy," *Financial Times*, September 19, 2005, http://www.ft.com/cms/s/1/97db5d0e-293f-11da-8a5e-00000e2511c8.html.

1. TURBULENCE AND HARMONY

1. The White House, "Introduction to Outer Space," U.S. Government Printing Office, Washington, D.C., March 26, 1958, http://www.fas.org/spp/guide/usa/intro1958.html; Wikipedia, "Where no man has gone before," http://en.wikipedia.org/wiki/Where_no_man_has_gone_before.
2. "John F. Kennedy Moon Speech," Rice Stadium, September 12, 1962, http://er.jsc.nasa.gov/seh/ricetalk.htm.
3. Kris Kitto, "Q&A with Jon Meacham, Author, *Thomas Jefferson: The Art of Power*," *The Hill*, November 13, 2012, http://thehill.com/capital-living/cover-stories/267719-qaa-with-jon-meacham-author-thomas-jefferson-the-art-of-power-.
4. Niall Ferguson, *Civilization: The West and the Rest* (New York: The Penguin Press, 2011).
5. David Ward, compiler, "Billy Graham Political Newspaper Ad Campaign: 'Vote for Biblical Values,'" *Deseret News*, November 1, 2012, http://www.deseretnews.com/article/865565802/Billy-Graham-ad-Vote-for-biblical-values.html?pg=all.
6. Robert D. Putnam and David E. Campbell, *American Grace: How Religion Divides and Unites Us* (New York: Simon & Schuster, 2010).
7. Mark Bowden, "David Petraeus's Winning Streak," *Vanity Fair*, March 30, 2010, http://www.vanityfair.com/politics/features/2010/05/petraeus-exclusive-201005.
8. Romans 14:12; N. D. Batra, "The US Is a Nation Divided for Good Reasons," *The Statesman*, September 17, 2008.
9. Evelyn Nieves, "Judges Ban Pledge of Allegiance from Schools, Citing 'Under God,'" *New York Times*, June 27, 2002.
10. *West Virginia State Board of Education, et al. v. Walter Barnette, et al.*, 319 U.S. 624 (1943).
11. *New York Times*, "Opinion: 'One Nation Under God,'" June 27, 2002, http://www.nytimes.com/2002/06/27/opinion/one-nation-under-god.html.
12. *New York Times*, "Excerpts from Federal Court Ruling on the Pledge of Allegiance," June 27, 2002, http://www.nytimes.com/2002/06/27/us/excerpts-from-federal-court-ruling-on-the-pledge-of-allegiance.html.
13. In *Elk Grove Unified School District v. Newdow*, No. 02-1624, the Court decided, albeit inconclusively, to block the excision of the phrase *under God*, from the Pledge of Allegiance.
14. "Falwell 'Sorry' for Mohammed Remark," BBC News, October 13, 2002, http://news.bbc.co.uk/2/hi/americas/2323897.stm; N. D. Batra, "Ask Larry Flynt What to Do with Jerry Falwell," *The Statesman*, October 23, 2002, Section: Perspective.
15. "Falwell Apologizes to Gays, Feminists, Lesbians," CNN.com, http://articles.cnn.com/2001-09-14/us/Falwell.apology_1_thomas-road-baptist-church-jerry-falwell-feminists?_s=PM:US.
16. Lawrie Mifflin, "Media Talk; Falwell Takes on the Teletubbies," *New York Times*, February 15, 1999, http://www.nytimes.com/1999/02/15/business/media-talk-falwell-takes-on-the-teletubbies.html.
17. Larry Flynt and Kenneth Ross, *An Unseemly Man: My Life as Pornographer, Pundit, and Social Outcast* (New York: Phoenix Books, 2008).
18. *Hustler Magazine v. Falwell*, 485 U.S. 46 (1988).
19. Ibid.
20. John Keats in *Ode on a Grecian Urn*:

> "Beauty is truth, truth beauty,—that is all
> Ye know on earth, and all ye need to know."

21. Lee C. Bollinger, *The Tolerant Society* (New York: Oxford University Press, 1988).
22. Mark Kennedy, "Eve Ensler Crusades for Women When Not Writing about Them," *MSN News*, November 21, 2012, http://news.msn.com/pop-culture/eve-ensler-crusades-for-

women-when-not-writing-about-them; N. D. Batra, "Of 'Girlie Men' and 'Bushwhooping' Women," *The Statesman* , July 28, 2004.

23. *Los Angeles Times*, "Schwarzenegger Promise to Blow Up Boxes Fizzled," December 28, 2010, http://losangeles.cbslocal.com/2010/12/28/schwarzenegger-promise-to-blow-up-boxes-fizzled/.

24. Peter Nicholas, "Gov. Criticizes Legislators as 'Girlie Men,'" *Los Angeles Times*, July 18, 2004, http://articles.latimes.com/2004/jul/18/local/me-arnold18; *San Francisco Chronicle*, "Schwarzenegger Deems Opponents 'Girlie-Men,'" July 18, 2004, http://www.sfgate.com/cgi-bin/article.cgi?file=/chronicle/archive/2004/07/18/MNGH57NKAF1.DTL.

25. Whoopi Goldberg in a Kerry-Edwards fund-raiser in 2004: "That's why I'm here tonight. Because I love bush. But someone's giving bush a bad name. Someone has tarnished name of 'bush.' Someone has waged war, someone has deliberately misled the country, someone has attempted to amend the constitution, all in the name of bush. The bush I know and cherish would never do such things. My bush is smarter than that. And if my bush is smarter than that, you can understand just how dumb I think that other bush is. And anyone who would wave to Stevie Wonder is not fully there. I will do whatever it takes to restore bush to its rightful place and that ain't in the White House. Vote your heart and mind and keep bush where it belongs." http://politicalhumor.about.com/b/2004/07/23/whoopi-goldbergs-crass-bush-joke.htm.

26. Martin Kasindorf, "Slim-Fast Drops Goldberg over Bush Puns," *USA Today*, July 14, 2004, http://www.usatoday.com/news/politicselections/2004-07-14-goldberg-slimfast_x.htm.

27. Thomas Fitzgerald, "Caustic Tone of Kerry Gala Draws Criticism; Stars Derided President Bush as a Liar and Thug. Kerry Aides Said They Spoke for Themselves," *The Philadelphia Inquirer*, July 10, 2004, http://articles.philly.com/2004-07-10/news/25371387_1_kerry-aides-john-kerry-kerry-campaign.

28. Ibid.

29. Ian Morris, *Why the West Rules—for Now: The Patterns of History, and What They Reveal About the Future* (New York: Farrar, Straus and Giroux, 2010).

30. Paul Kennedy, *The Rise and Fall of the Great Powers: Economic Change and Military Conflict from 1500 to 2000* (New York: Random House, 1987). Also see Henry R. Nau, "Why 'The Rise and Fall of the Great Powers' Was Wrong," *Review of International Studies* 27 (2001): 579–92, http://nationalism.org/library/science/ir/nau/nau-ris-2001-27-04.pdf.

31. Paul Kennedy, *Preparing for the Twenty First Century* (New York: Random House, 1993).

32. U.S. Department of Commerce, Bureau of Economic Analysis, "Current-dollar GDP—the market value of the nation's output of goods and services—increased 3.6 percent, or $134.6 billion, in the first quarter to a level of $15,454.0 billion. In the fourth quarter, current-dollar GDP increased 3.8 percent, or $143.3 billion," May 31, 2012, http://www.bea.gov/newsreleases/national/gdp/gdpnewsrelease.htm.

33. Ronald Reagan, "Remarks on East-West Relations at the Brandenburg Gate in West Berlin," June 12, 1987, http://www.reagan.utexas.edu/archives/speeches/1987/061287d.htm.

34. Niall Fergusson, *Civilization: The West and the Rest* (New York: The Penguin Press, 2011); "Complexity and Collapse," *Foreign Affairs*, March/April 2010, 18–32.

35. Ibid. It's rather shocking that Professor Ferguson should have used the image of a Victorian railway terminus because the Victorian Age was the age of slums, mass poverty, and misery for most British working classes, as captured in the novels of Charles Dickens, in spite of the fact that Queen Victoria had assembled the largest empire ever built by any country in the history of mankind.

36. Mimi Alford, *Once Upon a Secret: My Affair with President John F. Kennedy and Its Aftermath* (New York: Random House, 2012). Also see Stanton Peele, "The Top Seven Kennedy Sex Scandals: Kennedys' Reckless Sex," *Psychology Today*, May 21, 2008, http://www.psychologytoday.com/blog/addiction-in-society/200805/reckless-sex-and-power-iii-the-top-seven-kennedy-sex-scandals.

37. Rush Limbaugh, *The Way Things Ought to Be* (New York: Pocket Books, 1993). Noam Chomsky, *Occupy* (Westfield, NJ: Zuccotti Park Press, 2012).

38. John W. Dawson, "Causality in the Freedom–Growth Relationship," *European Journal of Political Economy*, 19 (2003): 479–95, http://www.appstate.edu/~dawsonjw/ejpefinal.pdf; Robert A. Lawson and J. R. Clark, "Examining the Hayek-Friedman Hypothesis on Economic and Political Freedom," http://ideas.repec.org/a/eee/jeborg/v74y2010i3p230-239.html.

39. Bloomberg Businessweek, "Speed Dial: Nitin Nohria," May 6, 2010, http://www.businessweek.com/magazine/content/10_20/b4178022137507.htm. Po Bronson and Ashley Merryman wrote in *Newsweek*, "The necessity of human ingenuity is undisputed. A recent IBM poll of 1,500 CEOs identified creativity as the No. 1 'leadership competency' of the future. Yet it's not just about sustaining our nation's economic growth. All around us are matters of national and international importance that are crying out for creative solutions, from saving the Gulf of Mexico to bringing peace to Afghanistan to delivering health care. Such solutions emerge from a healthy marketplace of ideas, sustained by a populace constantly contributing original ideas and receptive to the ideas of others." "The Creativity Crisis," July 10, 2010, http://www.newsweek.com/2010/07/10/the-creativity-crisis.html.

40. Donald E. Pease, *Theodor Seuss Geisel* (New York: Oxford University Press, 2010).

41. Andrew S. Grove, *Only the Paranoid Survive: How to Identify and Exploit the Crisis Points That Challenge Every Business* (New York: Doubleday Business, 1996), 3–4.

42. Michel Foucault, *Fearless Speech*, edited by Joseph Pearson (Cambridge, MA: The MIT Press, 2001). From Semiotext (e), http://semiotexte.com/?p=474:

> The expression "fearless speech" is a rough translation of the Greek parrhesia, which designates those who take a risk to tell the truth; the citizen who has the moral qualities required to speak the truth, even if it differs from what the majority of people believe and faces danger for speaking it. Parrhesia is a verbal activity in which a speaker expresses his personal relationship to truth through frankness instead of persuasion, truth instead of flattery, and moral duty instead of self-interest and moral apathy.

43. BBC News, "Buffett Warns on Investment 'Time Bomb,'" March 4, 2003, http://news.bbc.co.uk/2/hi/2817995.stm.

44. Timothy Sandefur, "Innovation," *The Concise Encyclopedia of Economics*, Library of Economics and Liberty, http://www.econlib.org/library/Enc/Innovation.html. Also see Knowledge@Wharton, "How Seemingly Irrelevant Ideas Lead to Breakthrough Innovation," January 30, 2013, http://knowledge.wharton.upenn.edu/article.cfm?articleid=3180.

45. Nike + iPod, http://www.apple.com/ipod/nike/.

46. Jack Goldsmith and Tim Wu, *Who Controls the Internet: Illusions of a Borderless World* (New York: Oxford University Press, 2006), 105–25; Joseph Menn, *All the Rave: The Rise and Fall of Shawn Fanning' s Napster* (New York: Crown Business, 2003).

47. Walter Isaacson, *Steve Jobs* (New York: Simon & Schuster, 2011).

48. Joseph A. Schumpeter, *Capitalism, Socialism and Democracy*, 3rd edition (New York: Harper Perennial, 1962).

49. Lee C. Bollinger, *The Tolerant Society* (New York: Oxford University Press, 1986), 173.

50. Tina Turner, "We Don't Need Another Hero," YouTube, http://www.youtube.com/watch?v=x8W4DEMUSeM&feature=fvwrel.

51. See Sedition Act, http://www.constitution.org/rf/sedition_1798.htm.

52. Harold L. Nelson and Dwight L. Teeter Jr., *Law of Mass Communications: Freedom and Control of Print and Broadcasting Medium* (New York: The Foundation Press, 1986), 41.

53. Horatio Shephard and Horace Greely are credited with starting the Penny Press with *Morning Post* in 1833.

54. David W. Blight, "The Civil War and Reconstruction Era, 1845–1877," *Academic Earth*, http://academicearth.org/courses/the-civil-war-and-reconstruction-era-1845-1877.

55. David M. Kennedy, *Over Here: The First World War and American Society* (New York: Oxford University Press, 2004), 80.

56. *Schenck v. United States*, 249 U.S. 47 (1919).

57. *Brandenburg v. Ohio*, 395 U.S. 444 (1969).

58. *Whitney v. California*, 274 U.S. 357 (1927).
59. Ibid.
60. *Brandenburg v. Ohio.*
61. *New York Times Co. v. Sullivan,* 376 U.S. 254 (1964).
62. Brian Stelter, "CNN Cancels 'In the Arena' with Eliot Spitzer," *New York Times*, July 6, 2011, http://mediadecoder.blogs.nytimes.com/2011/07/06/cnn-cancels-in-the-arena-with-eliot-spitzer/.
63. "We are the party of HELL NO!" Sarah Palin at a McCain campaign rally in 2008, http://www.youtube.com/watch?v=L6FN7SYdLWg.

These growing feathers pluck'd from Caesar's wing
Will make him fly an ordinary pitch,
Who else would soar above the view of men
And keep us all in servile fearfulness.
—*Julius Caesar*, Act 1, Scene 1.1

2. HOW FREE SPEECH STIMULATES INNOVATIVE BEHAVIOR

1. Francis Fukuyama, *Our Posthuman Future: Consequences of the Biotechnology Revolution* (New York: Farrar Straus & Giroux, 2002), 101.
2. The White House Office of the Press Secretary: Remarks of President Barack Obama—As Prepared for Delivery—Signing of Stem Cell Executive Order and Scientific Integrity Presidential Memorandum, Washington: The White House, March 9, 2009, http://www.whitehouse.gov/the_press_office/Remarks-of-the-President-As-Prepared-for-Delivery-Signing-of-Stem-Cell-Executive-Order-and-Scientific-Integrity-Presidential-Memorandum/.
3. Freeman Dyson, "Our Biotech Future," *The New York Review of Books*, July 19, 2007, http://www.nybooks.com/articles/archives/2007/jul/19/our-biotech-future/?pagination=false; Francis Fukuyama, *Our Posthuman Future*.
4. Varnee Murugan, "Embryonic Stem Cell Research: A Decade of Debate from Bush to Obama," *Yale Journal of Biology and Medicine* 82, no. 3 (September 2009): 101–03, http://www.ncbi.nlm.nih.gov/pmc/articles/PMC2744932/.
5. "*Roe v. Wade* and Beyond," Frontline, PBS, January 19, 2006, http://www.pbs.org/wgbh/pages/frontline/clinic/wars/cases.html.
6. *J.B. v M.B.*, http://embryo.asu.edu/pages/jb-v-mb-2001. "Court Says Woman Can Bar Embryos' Use," *New York Times*, August 15, 2001, http://www.nytimes.com/2001/08/15/nyregion/15EMBR.htm.
7. Ibid.
8. Shari Roan, "Infertility Researchers Find Success Using Frozen Eggs," *Los Angeles Times*, October 17, 1997, http://tech.mit.edu/V117/N51/infertility.51w.html. Also see Reproductive Biology Associates at http://rba-online.com/ivf/index.php?Home-.
9. Elizabeth Cohen, "Freezing Your Eggs—The Costs and Other Realities," CNN, October 6, 2011, http://www.cnn.com/2011/10/06/health/fertility-eggs-embryos-empowered-patient/index.html.
10. See Intracytoplasmic Sperm Injection (ICSI), http://www.asrm.org/topics/detail.aspx?id=1276.
11. "Maria Menounos' Plans for Her Embryos," *The Chart*, CNN Health, October 6, 2011, http://thechart.blogs.cnn.com/2011/10/06/maria-menounos-plans-for-her-embryos/.
12. "Golden Eggs," Carlene Hempel, Boston.com, June 25, 2006, http://www.boston.com/news/globe/magazine/articles/2006/06/25/golden_eggs/?page=full.
13. Paul Kennedy, *The Rise and Fall of Great Powers*: *Economic Change and Military Conflict from 1500 to 2000* (New York: Random House, 1987).
14. Sergio Pistoi, "Facing Your Genetic Destiny," *Scientific American*, February 18, 2002, http://www.scientificamerican.com/article.cfm?id=facing-your-genetic-desti-2002-02-18.

15. Jerome Groopman, "Ethics of Reproduction," *Wall Street Journal*, March 23, 2002, A23. Also see Paul Lauritzen, "Richer Views of the Ethics of Reproduction," http://www.questia.com/googleScholar.qst?docId=5002493440. Lauritzen writes, "Writing in the *Wall Street Journal* in March, Jerome Groopman discussed the recent case of the thirty-year-old woman who screened her IVF embryos for the early-onset Alzheimer's gene that she herself carries. Although the woman will likely succumb to Alzheimer's disease before her daughter reaches her tenth birthday, the girl at least will presumably be free of this disease."

16. The opening narration of *Star Trek* (1966).

17. Servicing Mission 4 (SM4) carried out by *Atlantis* was the last scheduled shuttle mission for the Hubble Space Telescope in May 2009.

18. Thomas Paine, *Common Sense* (1976) (New York: Forgotten Books, 2008), 61, http://www.forgottenbooks.org/index.php.

19. See Ilya Prigogine and Isabelle Stengers, *Order Out of Chaos: Man's New Dialogue with Nature* (New York: Bantam Books, 1984).

20. See *Joseph Burstyn, Inc. v. Wilson*, 343 U.S. 495 (1952) regarding the First Amendment protection of free speech in films; *Central Hudson Gas & Elec. Corp. v. Public Service Comm' n*, 447 U.S. 557 (1980) regarding censorship of advertising; *Jenkins v. Georgia*, 418 U.S. 153 (1974) regarding the conviction of a man for showing the film *Carnal Knowledge* in a movie theater; *Reno v. ACLU*, 521 US 844 (1997) concerning the Communications Decency Act; and *Eldred v. Ashcroft*, 537 U.S. 186 (2003) challenging the constitutionality of the 1998 Sonny Bono Copyright Term Extension Act (CTEA).

21. For discussion about innovation, see Lawrence Lessig, "The Architecture of Innovation," *Duke Law Journal* 51: 1783, http://www.law.duke.edu/shell/cite.pl?51+Duke+L.+J.+1783. Also see Paul Heckel, "Patents, Ecology and the First Amendment," http://www.pandab.org/patents-ecology-first-amendment.html.

22. For an authoritative account of the development of popular music, see Rachel Rubin and Jeffrey Paul Melnick, ed., *American Popular Music: New Approaches to the Twentieth Century* (Amherst, NY: The University of Massachusetts Press, 2001).

23. "Sputnik and the Dawn of the Space Age," NASA.gov, http://history.nasa.gov/sputnik/.

24. Regarding Nazi science, see Dennis Piszkiewicz, *The Nazi Rocketeers: Dreams of Space and Crimes of War* (New York: Praeger, 1995).

25. See Marius B. Jansen, *The Making of Modern Japan* (Cambridge, MA: Harvard University, 2000). Also, Williamson Murray and Allan Reed Millet, *A War to Be Won: Fighting the Second World War* (Cambridge: MA: Harvard University Press, 2000).

26. Robert O. Keohane, "Hegemony and After: Knowns and Unknowns in the Debate over Decline," *Foreign Affairs* 91, no. 4 (July–August 2012): 118.

27. Thomas Friedman, "The Secret of our Sauce," *New York Times*, March 7, 2004, http://www.nytimes.com/2004/03/07/opinion/the-secret-of-our-sauce.html. *The World Is Flat: A Brief History of the Twenty-First Century*, third edition (New York: Farrar, Straus and Giroux, 2007).

28. Fareed Zakaria, *The Post-American World* (New York: W. W. Norton, 2008).

29. In his well-acclaimed book, *The Wealth of Networks*, Yochai Benkler of Harvard University Law School discusses how the phenomenon of social production (social sharing and creation of knowledge) transforms markets.

30. It is important to understand the explanatory power of the theory of self-organizing systems as it applies to American society, where the unfettered and free marketplace of ideas under the First Amendment shapes how Americans conduct themselves publically. "The field of self-organization," according to Chris Lucas, "seeks general rules about the growth and evolution of systemic structure, the forms it might take, and finally methods that predict the future organization that will result from changes made to the underlying components. . . . The essence of self-organization is that system structure often appears without explicit pressure or involvement from outside the system. In other words, the constraints on form (i.e., organization) of interest to us are *internal to the system* (italics added), resulting from the interactions among the components and usually independent of the physical nature of those components." See Chris Lucas, Self-Organizing Systems (SOS) FAQ, Version 3, September 2008, http://www.calresco.org/sos/sosfaq.htm.

31. David E. Brown, *Inventing Modern America: From the Microwave to the Mouse* (Cambridge: MIT Press, 2002).

32. James Burke, *Circles: Fifty Round Trips through History Technology Science Culture* (New York: Simon and Schuster, 2009), 73.

33. Ashok Gadgil's profile available at http://web.mit.edu/invent/www/ima/gadgil_bio.html.

34. U.S. Patents Chart: Calendar Year 1963–2011, http://www.uspto.gov/web/offices/ac/ido/oeip/taf/us_stat.htm. Also see Daron Acemoglu, James A. Robinson, and Thierry Verdier, "Can't We All Be More Like Scandinavians?" MIT, March 2012, http://economics.mit.edu/files/8086.

35. For the origin of the Royal Society, see http://royalsociety.org/about-us/history/.

36. Royal Society, Wikipedia, http://en.wikipedia.org/wiki/Royal_Society#cite_note-35.

37. Goldie Blumenstyk, "Universities Report $1.8-Billion in Earnings on Inventions in 2011," *The Chronicle of Higher Education*, August 28, 2012, http://chronicle.com/article/University-Inventions-Earned/133972/.

38. Dahleen Glanton, "German Rocketeers' Role Praised: Their Expertise, a Legacy of WWII, Allowed U.S. to Win Space Race," *Chicago Tribune*, July 19, 1999, http://articles.chicagotribune.com/1999-07-19/news/9907190024_1_von-braun-space-travel-rocket-engineer-wernher-von.

39. Patrick Thibodeau, "IBM, Sun, HP and Intel Join on Grid Development," *Computerworld*, January 24, 2005.

40. Grid Computing Info Centre: Frequently Asked Questions (FAQ), http://www.gridcomputing.com/gridfaq.html.

41. The Writings of Benjamin Franklin: Philadelphia, 1726–1757, http://www.historycarper.com/resources/twobf2/advice.htm.

42. M. Mitchell Waldrop, "Grid Computing," *MIT Technology Review*, May 1, 2002, https://www.technologyreview.com/computing/12836/page4/.

43. Benjamin Schwarz, "Geography Is Destiny," *The Atlantic*, December 1, 2008, http://www.theatlantic.com/magazine/archive/2008/12/geography-is-destiny/307163/.

44. Mario Coccia, "Democratization Is the Determinant of Technological Change," National Research Council (Italy) and Max Planck Institute of Economics (Germany) CERIS-CNR, http://www.ceris.cnr.it/ceris/workingpaper/2008/WP_6_08_COCCIA.pdf.

45. Richard Florida, the author of *The Rise of Creative Class*, comes to similar conclusions. In "What's Happening to American Innovation?" in *The Atlantic*, he says, "The evidence here does not appear to indicate a significant innovation shortfall. The most we can say is that the rate of innovation has leveled off in recent years when we control for population. Nonetheless, the overall trajectory of American innovation remains consistently up . . . (but) the picture gets a bit more complicated when we parse patents by U.S.-born (resident) and foreign (non-resident) inventors." http://correspondents.theatlantic.com/richard_florida/2009/07/patent_trends_innovation_since_1980.php. Also see: "Global Sources of American Innovation," *The Atlantic*, July 15, 2009, http://www.theatlantic.com/national/archive/2009/07/global-sources-of-american-innovation/20571/.

46. Lawrence Lessig, "The Architecture of Innovation," Frey Lecture, Duke Law School, http://www.law.duke.edu/pd/papers/lessig.pdf.

47. Craig Newmark, "The Acceptance of Failure as a Spur to Innovation," Craigconnects, July 14, 2009, http://www.cnewmark.com/2009/07/the-acceptance-of-failure-as-a-spur-to-innovation.html.

3. TELLING STORIES TO A HYPERCONNECTED WORLD

1. *New York Times Co. v. United States*, 403 U.S. 713 (1971).

2. Lewis Friedland, *Public Journalism: Past and Future* (Dayton, OH: Kettering Foundation Press, 2003); Dan Gillmor, *We the Media: Grassroots Journalism by the People, for the People* (Sebastopol, CA: O'Reilly Media, 2004).

3. *New York Times v. Sullivan*, 376 U.S. 254 (1964).

4. PBS NewsHour, "*Times-Picayune* Editor on Commitment, Accountability Amidst Cutbacks," June 13, 2012, http://www.pbs.org/newshour/bb/media/jan-june12/nolapaper_06-13.html.

5. Stanley Fish, "Happy Birthday, Milton," *New York Times*, July 13, 2008, http://opinionator.blogs.nytimes.com/2008/07/13/happy-birthday-milton/.

6. Theodore Roosevelt, "The Man with the Muck-Rake," *The Art of Manliness*, April 14, 1906, http://artofmanliness.com/the-man-with-the-muck-rake-by-theodore-roosevelt/.

> There is filth on the floor, and it must be scraped up with the muck rake; and there are times and places where this service is the most needed of all the services that can be performed. But the man who never does anything else, who never thinks or speaks or writes, save of his feats with the muck rake, speedily becomes, not a help but one of the most potent forces for evil. There are in the body politic, economic and social, many and grave evils, and there is urgent necessity for the sternest war upon them. There should be relentless exposure of and attack upon every evil man, whether politician or business man, every evil practice, whether in politics, business, or social life. I hail as a benefactor every writer or speaker, every man who, on the platform or in a book, magazine, or newspaper, with merciless severity makes such attack, provided always that he in his turn remembers that the attack is of use only if it is absolutely truthful.

7. Leonard Downie Jr. and Michael Schudson, "The Reconstruction of American Journalism," *Columbia Journalism Review*, October 19, 2009, http://www.cjr.org/reconstruction/the_reconstruction_of_american.php?page=all.

8. Arnold J. Toynbee, "Challenge and Response," from *A Study of History*, http://www.cooperativeindividualism.org/toynbee_challenge_and_response.html.

> We have come to see that creation is the outcome of an encounter, that genesis is a product of interaction. . . . We shall be prepared now to recognize that, even if we were exactly acquainted with all the racial, environmental, and other data that are capable of being formulated scientifically, we should not be able to predict the outcome of the interaction between the forces which these data represent, any more than a military expert can predict the outcome of a battle or campaign from an "inside knowledge" of the dispositions and resources of both the opposing general staffs, or a bridge expert the outcome of a game from a similar knowledge of all the cards in every hand.

9. Committee to Protect Journalists, "CPJ's 2008 Prison Census: Online and in Jail," http://www.cpj.org/reports/2008/12/cpjs-2008-prison-census-online-and-in-jail.php.

10. Agents of Discovery, "A Report on the Incidence of Subpoenas Served on the News Media," http://www.rcfp.org/rcfp/orders/docs/AGENTS.pdf.

11. David E. Sanger, *Confront and Conceal: Obama's Secret Wars and Surprising Use of American Power* (New York: Crown Publishers, 2012).

12. The Reporter Committee on the Freedom of the Press, "Taricani Ordered Confined to Home on Criminal Contempt Charge," December 9, 2004, http://www.rcfp.org/browse-media-law-resources/news/taricani-ordered-confined-home-criminal-contempt-charge.

13. Jack Shafer, "The Times Scoops That Melted," *Slate*, July 25, 2003, http://www.slate.com/id/2086110/.

14. Carl Bernstein and Bob Woodward, *All the President's Men* (New York: Simon and Schuster, 1974).

15. Franklin Foer, "The Source of the Trouble," *New York Magazine*, May 21, 2005, http://nymag.com/nymetro/news/media/features/9226/.

16. Paul McMasters, "Journalists Can't Be Cops, Too," First Amendment Center, http://www.freedomforum.org/templates/document.asp?documentID=4578&printerfriendly=1.

17. John B. Zelezny, *Communications Law: Liberties, Restraints, and the Modern Media*, sixth edition (Belmont, CA: Thomson Wadsorth, 2011).

18. Zechariah Chaffee, *Free Speech in the United States* (Cambridge, MA: Harvard University Press, 1941), 31.

19. "The Watergate Story," *Washington Post*, http://www.washingtonpost.com/wp-srv/politics/special/watergate/#chapters.

20. Kurt Eichenwald, *Conspiracy of Fools: A True Story* (New York: Broadway Books, 2005); Robert Bryce, *Pipe Dreams: Greed, Ego, and the Death of Enron* (New York: PublicAffairs, 2002).

21. After putting up a hopeless defense, CBS asked former attorney general Dick Thornburgh and former Associated Press president Louis Boccardi to conduct an independent investigation into the Bush National Guard story. According to the panel report: "The combination of a new *60 Minutes Wednesday* management team, great deference given to a highly respected producer and the network's news anchor, competitive pressures, and a zealous belief in the truth of the segment seem to have led many to disregard some fundamental journalistic principles." PBS NewsHour, "CBS Ousts Four Executives over Disputed *60 Minutes* Report," January 10, 2005, http://www.pbs.org/newshour/updates/cbsnews_01-10-05.html. Jay Rosen, "After Trust Me Journalism Comes Openness: Rather Report Released," http://journalism.nyu.edu/pubzone/weblogs/pressthink/2005/01/10/cbs_rept05.html; Howard Kurtz, "Dan Rather to Step Down at CBS: Anchor's Decision Comes Amid Probe of Flawed Bush Report," *Washington Post*, November 24, 2004, A, http://www.washingtonpost.com/wp-dyn/articles/A7313-2004Nov23.html.

22. Michael Dobbs and Mike Allen, "Some Question Authenticity of Papers on Bush," *Washington Post*, September 10, 2004, A01, http://www.washingtonpost.com/wp-dyn/articles/A9967-2004Sep9.html.

23. *King Lear*, Act III, Scene 2.

24. Adam Cohen, "The Press and the Dress," *Time*, February 2, 1998, http://www.cnn.com/ALLPOLITICS/1998/02/09/time/cohen.html.

25. Nelson Johnson, *Boardwalk Empire: The Birth, High Times, and Corruption of Atlantic City* (Medford, NJ: Plexus Publishing Company, 2002).

26. Preeti Bhattacharji, "Uighurs and China's Xinjiang Region," Council on Foreign Relations, May 29, 2012, http://www.cfr.org/china/uighurs-chinas-xinjiang-region/p16870#p4.

27. Brian Whitaker and Luke Harding, "American Beheaded in Revenge for Torture," *The Guardian*, May 12, 2004, http://www.guardian.co.uk/world/2004/may/12/iraq.alqaida.

28. "Editor Sacked over 'Hoax' Photos," BBC, May 14, 2004, http://news.bbc.co.uk/2/hi/uk_news/politics/3716151.stm.

29. Dan Rather and Mickey Herskowitz, *The Camera Never Blinks: Adventures of a TV Journalist* (New York: William Morrow, 1977).

30. Margie Mason, Associated Press, "Iconic 'Napalm Girl' Photo from the Vietnam War Turns 40," USA Today, June 2, 2012, http://www.usatoday.com/news/world/story/2012-06-02/napalm-girl-photo-vietnam/55347678/1; Eddie Adams, "Murder of a Vietcong by Saigon Police Chief – Eddie Adams , 1968," http://www.pbase.com/omoses/image/118045027.

31. Katherine Q. Seelye, "Newsweek Apologizes for Report of Koran Insult," *New York Times*, May 16, 2005, http://www.nytimes.com/2005/05/16/international/asia/16koran.html?pagewanted=all.

32. Seymour M. Hersh, *Chain of Command: The Road from 9/11 to Abu Ghraib* (New York: HarperCollins, 2004). Also see Seymour M. Hersh, "Chain of Command: How the Department of Defense Mishandled the Disaster at Abu Ghraib," *The New Yorker*, May 17, 2004, http://www.newyorker.com/archive/2004/05/17/040517fa_fact2#ixzz1y9DNSZy3.

33. Ted Kennedy quotes, http://www.qotd.org/search/search.html?aid=1865&page=2. Also see *Congressional Records-Senate*, June 4, 2004, http://books.google.com/books?id=ZMdG0AlVQZoC&pg=PA11548&lpg=PA11548&dq.

34. Daniel Okrent, "Weapons of Mass Destruction? Or Mass Distraction?" *New York Times*, May 30, 2004, http://www.nytimes.com/2004/05/30/weekinreview/30bott.html. Howard Kurtz, "The Post on WMDs: An Inside Story," *Washington Post*, August 12, 2004, http://www.washingtonpost.com/wp-dyn/articles/A58127-2004Aug11.html.

35. Associated Press, "Web Site Shows Apparent Beheading Tape," *USA Today*, August 11, 2008, http://www.usatoday.com/news/world/iraq/2004-08-11-beheading_x.htm.

36. Matthew B. Stannard, "Beheading Video Seen as War Tactic," *San Francisco Chronicle*, May 13, 2004, http://www.sfgate.com/cgi-bin/article.cgi?file=/c/a/2004/05/13/MNG6E6KL791.DTL.

37. The FCC's prohibition against the broadcast of hoaxes is set forth at Section 73.1217 of the Commission's rules, 47 C.F.R. § 73.1217.

38. James Dao and Eric Schmitt, "A Nation Challenged: Hearts and Minds; Pentagon Readies Efforts to Sway Sentiment Abroad," *New York Times*, February 19, 2002, http://www.nytimes.com/2002/02/19/world/nation-challenged-hearts-minds-pentagon-readies-efforts-sway-sentiment-abroad.html?pagewanted=all&src=pm.

39. For an account of collusion between the CIA and American news media see Carl Bernstein in a *Rolling Stone* magazine article from 1977 titled "The CIA and the Media," http://www.carlbernstein.com/magazine_cia_and_media.php.

40. See, for example, C. Edwin Baker, *Media Concentration and Democracy: Why Ownership Matters* (New York: Cambridge University Press, 2006).

41. Russell Adams and Shira Ovide, "Newspapers Move to Outsource Foreign Coverage," *Wall Street Journal*, January 15, 2009, http://online.wsj.com/article/SB123197973917183829.html.

42. Mike Allen and Michael Calderone, "*Washington Post* Cancels Lobbyist Event Amid Uproar," *Poltico*, July 3, 2009, http://www.politico.com/news/stories/0709/24441.html.

43. YouTube Reporters' Center, http://www.youtube.com/user/reporterscenter.

44. Ibid.

45. Edward Wasserman, "Commentary: Accuracy Standards and Online Journalism," *Miami Herald*, June 23, 2009, http://www.mcclatchydc.com/2009/06/23/v-print/70475/commentary-accuracy-standards.html.

46. Jayson Blair was a *New York Times* journalist who was fired in 2003 for his fake reporting. See "Correcting The Record; Times Reporter Who Resigned Leaves Long Trail of Deception," *New York Times*, November 5, 2003, http://www.nytimes.com/2003/05/11/us/correcting-the-record-times-reporter-who-resigned-leaves-long-trail-of-deception.html?pagewanted=all&src=pm.

47. Ibid.

48. Ibid.

49. Chris Anderson, *The Long Tail: Why the Future of Business Is Selling Less of More* (New York: Hyperion, 2006).

50. Edward Wasserman, "Threats to Ethical Journalism in the New Media Age," *Media Ethics Online* 21, no. 2 (Spring 2010), http://www.mediaethicsmagazine.com/index.php/browse-back-issues/132-spring-2010/3919175-threats-to-ethical-journalism-in-the-new-media-age.

51. N. D. Batra, "Message Is Loud and Clear," *The Times of India*, September 4, 2009, http://timesofindia.indiatimes.com/home/opinion/edit-page/Top-Article-Message-Is-Loud-And-Clear/articleshow/4968478.cms.

52. Marshall McLuhan, *Understanding Media: The Extensions of Man* (London: Routledge, 1964).

53. Janis Krums, Twitpic, http://twitpic.com/135xa.

54. Jacqui Cheng, "Landlord Sues Tenant after Tweet about Moldy Apartment," *Arts Technica*, July 28, 2009, http://arstechnica.com/tech-policy/news/2009/07/landlord-sues-tenant-after-tweet-about-moldy-apartment.ars.

55. Eriq Gardner, "Courtney Love Sued for Twitter Defamation," *Reuters*, March 30, 2009, http://www.reuters.com/article/idUSTRE52R00020090331.

56. WJS Blogs, "Obama's Twitter Q&A on Fiscal Cliff," *Wall Street Journal*, December 3, 2012, http://blogs.wsj.com/washwire/2012/12/03/obamas-twitter-qa-on-fiscal-cliff/?mod=e2tw.

57. Steven R. Weisman, "U.S. Said to Compromise on Beef for South Korea," *New York Times*, June 20, 2008, http://www.nytimes.com/2008/06/20/washington/20trade.html.

58. Digital Life, "Discover How the World Lives Online," TNS Digital Life, http://discoverdigitallife.com/.

59. Quoted in *DirectNews*, http://www.directnews.co.uk/news/online-marketing/social-networking/social-networking-will-be-depended-on-in-the-future-$21377860.htm.

60. Mark S. Granovetter, "The Strength of Weak Ties," *American Journal of Sociology* 78, no. 6 (May 1973): 1360–80, http://www.stanford.edu/dept/soc/people/mgranovetter/documents/granstrengthweakties.pdf.

61. Michael Clarke, the executive vice president for Product and Market Development at Silverchair Information Systems, wrote in the *Scholarly Kitchen*: "The structure of Twitter's open, content-centric network enables information diffusion via weak ties." See "The Strength of Weak Ties: Why Twitter Matters in Scholarly Communication," *The Scholarly Kitchen*, August 12, 2009, http://scholarlykitchen.sspnet.org/2009/08/12/the-strength-of-weak-ties-why-twitter-matters-in-scholarly-communication/. Also see Sara Grant, "Caves, Clusters, and Weak Ties: The Six Degrees World of Inventors," Q&A with Lee Fleming, Working Knowledge, Harvard Business School, November 9, 2004, http://hbswk.hbs.edu/item/4516.html.

62. Andreas Cremer, "Porsche Blocks Staff Access to Facebook as Espionage Shield," *Bloomberg*, October 9, 2010, http://www.bloomberg.com/news/2010-10-09/porsche-blocks-staff-access-to-facebook-as-espionage-shield-wiwo-reports.html.

63. Clay Shirky, *Cognitive Surplus: Creativity and Generosity in a Connected Age* (New York: The Penguin Press, 2010), 25.

64. Jared Keller, "Predicting Stock Market Changes Using Twitter," *The Atlantic*, October 20, 2010, http://www.theatlantic.com/technology/archive/2010/10/predicting-stock-market-changes-using-twitter/64897/.

65. Ben Adler, "A Look at Tea Party Election-Day Battle Plans," *Newsweek*, October 22, 2010, http://www.newsweek.com/2010/10/22/a-look-at-tea-party-election-day-battle-plans.html?from=rss.

66. Contrary to what Peter Steiner said, "On the Internet,Nobody Knows You're a Dog," *The New Yorker*, July 5, 1993, http://www.unc.edu/depts/jomc/academics/dri/idog.html.

67. Eric Schmidt, "Welcome to 'Age of Augmented Humanity,'" http://vodpod.com/watch/5445636-eric-schmidts-talk-on-age-of-augmented-humanity-at-dld-video.

68. Lawrence Lessig, http://lessig.org/.

69. J. B. Walther, "Computer-Mediated Communication: Impersonal, Interpersonal, and Hyperpersonal Interaction," *Communication Research* 23 (1996): 3–43.

70. Tim Craig and Michael D. Shear, "Allen Quip Provokes Outrage, Apology," *Washington Post*, August 15, 2006, http://www.washingtonpost.com/wp-dyn/content/article/2006/08/14/AR2006081400589.html.

71. AFP, "US Senate Hopeful Regrets Telling Voters 'I'm Not a Witch,'" October 21, 2010, http://www.google.com/hostednews/afp/article/ALeqM5hF0ID4s0zt08u-ioW_v1kfZzOQ3g?docId=CNG.333b518801c245da982405e79dfd3d93.141 .

72. Adults and Social Network Websites, Pew Internet, January 14, 2019, http://www.pewinternet.org/Reports/2009/Adults-and-Social-Network-Websites/2-Main-Findings/5-More-than-half-of-adult-social-network-users-have-more-than-one-profile-online.aspx.

73. N. D. Batra, *Digital Freedom, How Much Can You Handle?* (New York: Rowman and Littlefield, 2007).

74. Pew Research Center, "Teens and Social Media," Pew Internet and Life Project, December 19, 2008, http://www.pewinternet.org/Reports/2007/Teens-and-Social-Media.aspx.

75. Robert D. Putnam, *Bowling Alone: The Collapse and Revival of American Community* (New York: Touchstone Books, 2001).

76. Malcolm Gladwell, "Small Change: Why Revolution Will Not Be Tweeted," *The New Yorker*, October 4, 2010, http://www.newyorker.com/reporting/2010/10/04/101004fa_fact_gladwell?currentPage=all .

77. Ibid.

78. Emily McClintock Ekins and David Kirb, "Libertarian Roots of the Tea Party," Cato Institute, August 6, 2012, http://www.cato.org/publications/policy-analysis/libertarian-roots-tea-party?utm_source=Cato+Institute+Emails&utm_campaign=6b9d637298-Research_and_Analysis&utm_medium=email&mc_cid=6b9d637298&mc_eid=a1708a475b.

William Yardley, "The Branding of the Occupy Movement," *New York Times* , November 27, 2011, http://www.nytimes.com/2011/11/28/business/media/the-branding-of-the-occupy-movement.html?pagewanted=all.

79. Benjamin Franklin, *Poor Richard' s Almanack*, 1735.

80. Stephan Faris, "The Vatican Scandals: What Did the Pope's Butler Know?" *Time*, May 29, 2012, http://www.time.com/time/world/article/0,8599,2115967,00.html#ixzz1yLdDWkjI.

81. "Roman Catholic Church Sex Abuse Cases," *New York Times*, March 12, 2012, http://topics.nytimes.com/top/reference/timestopics/organizations/r/roman_catholic_church_sex_abuse_cases/index.html.

82. Presidential Apology, The White House, May 16, 1997, http://www.cdc.gov/tuskegee/clintonp.htm.

83. Mike Wallace, "1996: Tobacco and Wigand," *60 Minutes*, CBS, February 4, 1996, http://www.cbsnews.com/video/watch/?id=4455582n.

84. For a list of whistleblowers, see http://en.wikipedia.org/wiki/List_of_whistleblowers#.

85. Elizabeth Dickinson, "The First WikiLeaks Revolution?" *Foreign Policy*, January 13, 2011, http://wikileaks.foreignpolicy.com/posts/2011/01/13/wikileaks_and_the_tunisia_protests.

86. Amy Goodman, "Collateral Murder in Iraq," *TruthDig*, April 6, 2010, http://www.truthdig.com/report/m/collateral_murder_in_iraq_20100406/.

87. Timothy Garton Ash, "US Embassy Cables: A Banquet of Secrets," *The Guardian*, November 28, 2010, http://www.guardian.co.uk/commentisfree/2010/nov/28/wikileaks-diplomacy-us-media-war.

88. Tom Chivers, "WikiLeaks' 10 Greatest Stories," *The Telegraph*, October 18, 2010, http://www.telegraph.co.uk/news/worldnews/asia/afghanistan/8070253/Wikileaks-10-greatest-stories.html.

89. Elizabeth Bumiller, "Gates on Leaks, Wiki and Otherwise," *New York Times*, November 30, 2010, http://thecaucus.blogs.nytimes.com/2010/11/30/gates-on-leaks-wiki-and-otherwise/. Forty years ago, Justice Potter Stewart, in the Pentagon Paper case decision, expressed similar opinion. He wrote:

> But I cannot say that disclosure of any of them will surely result in direct, immediate, and irreparable damage to our Nation or its people. . . . It is elementary that the successful conduct of international diplomacy and the maintenance of an effective national defense require both confidentiality and secrecy. Other nations can hardly deal with this Nation in an atmosphere of mutual trust unless they can be assured that their confidences will be kept. . . . And within our own executive departments, the development of considered and intelligent international policies would be impossible if those charged with their formulation could not communicate with each other freely, frankly, and in confidence . . . The Executive has relatively unchecked powers, therefore, a free press is all the more important as the only foundation for a critical and enlightened public that could act as a check on abuse of Executive power. . . . The executive has the responsibility to make sure that only what needs classifying is indeed classified, and that which is classified does not get leaked . . . Consistent with the First Amendment, the court cannot stop the publication of leaked materials. . . . That messy balance between the administration's need for secrecy and the public's right and need to know, while far from perfect, does mean that the administration continues to function under normal conditions, subject to occasional disclosures to keep it honest. . . . The rare exceptions would require a combination of high likelihood, magnitude, and immediacy of harm to justify suppression. In the area of national defense, this is captured by the phrase "the sailing dates of transports or the number and location of troops."—*New York Times Co. v. United States*, 403 U.S. 713 (1971)

90. Maxwell E. McCombs and Donald L. Shaw, The Agenda-Setting Function of Mass Media," *Public Opinion Quarterly* 36, no. 2: 176–87, http://poq.oxfordjournals.org/content/36/2/176.short.

91. Jason Pontin, "Organizations care about those who pay their bills: if a media company's revenues derive solely from advertising, then it will be advertisers whom publishers ask editors to serve. If the editorial published by media companies is not to be deformed in subtle and not-so-subtle ways, readers must have some skin in the game." "*Technology Review* Goes Digital First," *Technology Review*, June 4, 2012, http://www.technologyreview.com/view/428072/technology-review-goes-digital-first/.

92. Peter G. Goheen, "The Impact of the Telegraph on the Newspaper in Mid-Nineteenth Century British North America," *Urban Geography* 11 (1990), http://bellwether.metapress.com/content/t3114w344855723n/?referencesMode=Show. Menahem Blondheim, *News over the Wires: The Telegraph and the Flow of Public Information in America, 1844–1897* (Cambridge, MA: Harvard University Press, 1994).

93. B. Kovarik, *Revolutions in Communication: Media History from Gutenberg to the Digital Age* (New York: The Continuum International Publishing Group, 2011).

94. Jason Pontin, "How to Save Media," *Technology Review*, May 4, 2009, http://www.technologyreview.com/view/413355/how-to-save-media/.

95. Michael Schudson, *The Sociology of News*, second edition (New York: W. W. Norton & Company, 2011), 207.

4. COMMERCIAL SPEECH AND INNOVATION

1. Amar Bhidé, "Venturesome Consumption, Innovation and Globalization," presented in Venice at the CESifo and Centre on Capitalism and Society conference, July 21–22, 2006, http://www.bhide.net/bhide_ venturesome _ consumption .pdf.

2. *Virginia State Board of Pharmacy v. Virginia Citizens Consumer Council, Inc.*, 425 U.S. 748 (1976).

3. Clayton M. Christensen, *The Innovator's Dilemma: When New Technologies Cause Great Firms to Fail* (Boston, MA: Harvard Business School Press, 1997); Clayton M. Christensen and Michael E. Raynor, *The Innovator's Solution: Creating and Sustaining Successful Growth* (Boston, MA: Harvard Business Press, 2003).

4. Dave Itzkoff, "HBO Picks Up Aaron Sorkin Series About Cable TV News," *New York Times*, September 8, 2011, http://artsbeat.blogs.nytimes.com/2011/09/08/hbo-picks-up-aaron-sorkin-series-about-cable-tv-news/.

5. *Valentine v. Chrestensen*, 316 U.S. 52 (1942); Alex Kozinski and Stuart Banner, "Who's Afraid of Commercial Speech?" *Virginia Law Review* 76 (May 1990): 627, http://notabug.com/kozinski/fearcommerce.

6. *Bigelow v. Virginia*, 421 U.S. 809 (1975), http://www.law.cornell.edu/supct/html/historics/USSC_CR_0421_0809_ZS.html; Also, see *Roe v. Wade*, 410 U.S. 113 (1973), and *Doe v. Bolton*, 410 U.S. 179 (1973).

7. Steven Johnson, *Where Good Ideas Come From: The Natural History of Innovation* (New York: Riverhead Publisher, 2010).

8. *Virginia State Pharmacy Board v. Virginia Citizens Consumer Council*, 425 U.S. 748, 753–54 (1976).

9. Ibid.

10. Ibid.

11. *Central Hudson Gas & Elec. Corp. v. Public Serv. Comm'n*, 447 U.S. 557 (1980).

12. Ibid.

13. *Board of Trustees of the State Univ. of N.Y. v. Fox*, 492 U.S. 469, 482 (1989).

14. E. S. Turner, *The Shocking History of Advertising*, Revised Edition (New York: Penguin Books, 1965), 9.

15. Biographies, "Bill Rancic," http://www.theapprenticerules.com/biographies/profile-bill-rancic.html.

16. In the Matter of Trump Hotels & Casino Resorts, Inc., January 16, 2002, http://www.sec.gov/litigation/admin/34-45287.htm.

17. Leigh Dyer, "Charlotte, N.C., Son Set to Reap Rewards of Reality-TV Run on the 'Apprentice,'" *Charlotte Observer*, N. C. Knight Ridder/Tribune Business News, April 17, 2004, http://www.highbeam.com/doc/1G1-115449382.html; *USA Today*, "Kwame Jackson Capitalizes on Being No. 2," April 20, 2004, http://usatoday30.usatoday.com/life/television/news/2004-04-20-apprentice-kwame_x.htm.

18. Anil K. Gupta and Haiyan Wang, "Chinese Innovation Is a Paper Tiger," *Wall Street Journal*, July 28, 2011, http://online.wsj.com/article/SB10001424053111904800304576472034085730262.html.

19. Vance Packard, *The Hidden Persuaders* (Brooklyn: Ig Publishing, 1957). Dave Lakhani, *Subliminal Persuasion: Influence & Marketing Secrets They Don't Want You to Know* (Hoboken, NJ: John Wiley & Sons, 2008); N. D. Batra, "Torn between KFC & Victoria's Secret," *The Statesman*, November 26, 2003.

20. David L. Katz, "The Scarlet Burger," *Wall Street Journal*, November 9, 2003.

21. In the Matter of KFC Corporation, Complaint, http://www.ftc.gov/os/caselist/0423033/040603comp0423033.pdf.

22. Michael J. Pelgro, "The Authority of the Federal Trade Commission to Order Corrective Advertising," *Boston College Law Review* 19, no 5 (July 1, 1978).

23. Kate Macarthur, "KFC Pulls Controversial Health-Claim Chicken Ads," *Ad Age*, November 18, 2003, http://adage.com/article/news/kfc-pulls-controversial-health-claim-chicken-ads/38857/; "Ad Age Editorial: KFC Blunders in 'Health' Ads," *Ad Age*, November 3, 2003, http://adage.com/results?endeca=1&searchprop=AdAgeAll&return=endeca&search_offset=0&search_order_by=score&search_phrase=KFC+BLUNDERS+IN+%27HEALTH%27+ADS.

24. V. I. Lenin, *The State and Revolution* (1917), Quoted in International League, http://www.icl-fi.org/print/english/wv/940/qotw.html.

25. Brooke Shields in the Calvin Klein Jeans commercial in 1981, http://www.youtube.com/watch?v=YK2VZgJ4AoM.

26. W. Huitt, "Maslow's Hierarchy of Needs," *Educational Psychology Interactive*, Valdosta, GA: Valdosta State University. Retrieved July 20, 2011, http://www.edpsycinteractive.org/topics/regsys/maslow.html; Abraham Maslow, *Motivation and Personality*, third edition (New York: Harper and Row Publishers, 1987).

27. A syndicated television game show that aired from 1991 to 2006.

28. Susan J. Carroll and Richard Logan Fox, *Gender and Elections: Shaping the Future of American Politics* (New York: Cambridge University Press, 2006), 93–94.

29. George A. Kennedy, *Aristotle, on Rhetoric: A Theory of Civic Discourse* (New York: Oxford University Press, 1991).

30. Robert A. Sevier, "Marketing," University Business, March 2006, http://www.universitybusiness.com/viewarticle.aspx?articleid=69&p=2.

31. Keith Bradsher, "India Slips Far Behind China, Once Its Close Economic Rival," *New York Times*, November 29, 2002, http://www.nytimes.com/2002/11/29/world/india-slips-far-behind-china-once-its-close-economic-rival.html.

32. Alissa Quart, "A Smarter Way to Sell Ketchup," *Wired*, December 1, 2002, http://www.wired.com/wired/archive/10.12/start.html?pg=2.

33. "Vivius to License First Patent for Pioneering Insurance Technology; Insurance Sectors beyond Consumer-Choice Healthcare Market to Benefit from Key Vivius Technology," *Business Wire*, September 13, 2004, http://www.businesswire.com/news/home/20040913005986/en/Vivius-License-Patent-Pioneering-Insurance-Technology-Insurance.

34. Jeffrey R. Kuester, "Congress Takes Another Aim at Business Method Patents," http://www.kuesterlaw.com/hr5364.htm.

35. Bronwyn H. Hall, "Business Method Patents, Innovation, and Policy," http://emlab.berkeley.edu/~bhhall/papers/BHH%20on%20BMP%20May03WP.pdf.

36. Jordan Weissmann, "The Case for Abolishing Patents (Yes, All of Them)," *The Atlantic*, September 27, 2012, http://www.theatlantic.com/business/archive/2012/09/the-case-for-abolishing-patents-yes-all-of-them/262913/.

37. *State Street Bank & Trust Co. v. Signature Financial Group Inc*., U.S. Court of Appeals Federal Circuit, July 23, 1998, 149 F.3d 1368, 47 USPQ2d 1596, http://digital-law-online.info/cases/47PQ2D1596.htm.

38. John Schwartz, "Justices Take Broad View of Business Method Patents," *New York Times*, June 28, 2010, http://www.nytimes.com/2010/06/29/business/29patent.html.

39. Adam Liptak, "Justices, 5–4, Reject Corporate Spending Limit," *New York Times*, January 21, 2010, http://www.nytimes.com/2010/01/22/us/politics/22scotus.html; Floyd Abrams, "The First Amendment Is Just Fine as Is," *New York Times*, October 24, 2012, http://www.nytimes.com/roomfordebate/2012/10/24/amend-the-constitution-to-limit-political-spending/the-first-amendment-is-just-fine-as-is.

5. FREE PRESS AND MARKETPLACE FUNDAMENTALISM

1. IMDb, "Memorable quotes for *Wall Street* (1987), http://www.imdb.com/title/tt0094291/quotes.

2. George W. Bush's "Remarks at the Corporate Fraud Conference," September 26, 2002, http://www.gpo.gov/fdsys/pkg/WCPD-2002-09-30/pdf/WCPD-2002-09-30-Pg1626.pdf.

3. Xiaoding Liu and Jay R. Ritter, "Corporate Executive Bribery: An Empirical Analysis," EVI conference (2006), Babson College, December 4, 2007, http://bear.warrington.ufl.edu/ritter/work_papers/BriberyDec42007.pdf; Sean J. Griffith, "Spinning and Underpricing: A Legal and Economic Analysis of the Preferential Allocation of Shares in Initial Public Offerings," *Brooklyn Law Review* 69, no. 2 (2004): 583–649, http://papers.ssrn.com/sol3/papers.cfm?abstract_id=525722.

4. Joe Nocera, "Rigging the I.P.O. Game," *New York Times*, March 9, 2013, http://www.nytimes.com/2013/03/10/opinion/sunday/nocera-rigging-the-ipo-game.html?_r=1&pagewanted=all&.

5. Dan Reingold and Jennifer Reingold, *Confessions of a Wall Street Analyst: A True Story of Inside Information and Corruption in the Stock Market* (New York: Harper Collins, 2006).

6. Daniel Kadlec, "Enron: Who's Accountable?" *Time*, January 13, 2002.

7. Milton Friedman, "The Social Responsibility of Business Is to Increase Its Profits," *New York Times Magazine*, September 13, 1970, http://www.colorado.edu/studentgroups/libertarians/issues/friedman-soc-resp-business.html.

8. Aymo Brunetti and Beatrice Weder, "A Free Press Is Bad News for Corruption," *Journal of Public Economics* 87 (2003): 1801–24; Sebastian Freille, M. Emranul Haque, and Richard Kneller, "A Contribution to the Empirics of Press Freedom and Corruption," *European Journal of Political Economy* 23 (2007): 838–62, http://personalpages.manchester.ac.uk/staff/emranul.haque/ejpe_%20press%20freed%20and%20corruption_dec2007.pdf.

9. Cristina Romero, "Reporting the Financial Crisis: A Media Failure?" European Journalism Center, November 16, 2012, http://www.ejc.net/magazine/article/Reporting_the_financial_crisis_a_media_failure/.

10. Leslie Stahl, *60 Minutes* Interview with Jack Abramoff, "Jack Abramoff: The Lobbyists Playbook," Transcript, November 6, 2011, http://www.cbsnews.com/2102-18560_162-57319075.html?tag=contentMain;contentBody.

11. "Bush: Don't Do as I Did," CNN Money, July 11, 2002, http://money.cnn.com/2002/07/11/news/bush_loan/.

12. *Enron and World Finance: A Case Study in Ethics*, ed. Paul H. Dembinski, Carole Lager, Andrew Cornford, and Jean-Michel Bonvin (New York: Palgrave Macmillan, 2006).

13. Elisabeth Bumiller, "Corporate Conduct: The President; Bush Signs Bill Aimed at Fraud in Corporations," *New York Times*, July 31, 2002, http://www.nytimes.com/2002/07/31/business/corporate-conduct-the-president-bush-signs-bill-aimed-at-fraud-in-corporations.html?pagewanted=all&src=pm.

14. The Federalist Society for Law and Public Policy Studies, "President Bush Signs Corporate Corruption Bill," July 30, 2002, http://www.fed- soc.org/publications/page/president-bush-signs-corporate-corruption-bill.

15. Jack Welch, *Jack: Straight from the Gut* (New York: Business Plus, 2001).
16. Pam Lambert, "X-Treme Divorce," People.com, October 14, 2002, www.people.com/people/archive/article/0,,20138217,00.html.
17. Geraldine Fabrikant and David Cay Johnston, "G.E. Perks Raise Issues About Taxes," *New York Times*, September 9, 2002, http://www.nytimes.com/2002/09/09/business/ge-perks-raise-issues-about-taxes.html.
18. "Stand by Your Man," a 1968 song coauthored by Tammy Wynette and Billy Sherrill.
19. Seneca County Ohio Court Records Directory, http://www.courtreference.com/Seneca-County-Ohio-Courts.htm.
20. Robert Mankoff, "Good news, chief, a computer virus destroyed all our documents," *The New Yorker*, January 28, 2002, http://www.condenaststore.com/-sp/Good-news-chief-a-computer-virus-destroyed-all-our-documents-New-Yorker-Cartoon-Prints_i8544749_.htm.
21. Paul R. La Monica, "Fiorina Out, HP Stock Soars," CNN/Money, February 10, 2005, http://money.cnn.com/2005/02/09/technology/hp_fiorina/.
22. AFX News, "HP to Pay Fiorina $21 Million Severance Package," February 14, 2005, http://web.archive.org/web/20050218232358/http://www.ecommercetimes.com/story/40594.html.
23. President George W. Bush, "Fact Sheet: America's Ownership Society: Expanding Opportunities," The White House, June 17, 2004, http://georgewbush-whitehouse.archives.gov/news/releases/2004/08/20040809-9.html.
24. Dan Ackman , "The WorldCom We Hardly Knew, " Forbes.com, June 26, 2002, http://www.forbes.com/2002/06/26/0626topnews.html.
25. PBS NewsHour, "Bernard Ebbers Guilty of Conspiracy and Fraud," Transcript, March 15, 2005, http://www.pbs.org/newshour/bb/business/jan-june05/ebbers_3-15.html.
26. Robert J. Shiller, "How Wall Street Learns to Look the Other Way," *New York Times*, February 8, 2005, http://www.nytimes.com/2005/02/08/opinion/08shiller.html?pagewanted=print&positio= .
27. Arthur Levitt Jr., "Money, Money, Money," *Wall Street Journal*, November 22, 2004, A14, http://www.cb.wsu.edu/~nwalcott/finance325/template/readings/Money,%20Money,%20Money.pdf.
28. Ibid.
29. Benjamin Wallace, "The Comeback That Wasn't," *New York Magazine*, July 31, 2011, http://nymag.com/news/features/martha-stewart-2011-8/. For insider trading cases, see Bloomberg.com, "Inside Trading News," http://topics.bloomberg.com/insider-trading/.
30. Clifford Coonan, "Singapore Oil Chief to Face Police over $550m Trade Losses," *Times*, December 7, 2004, http://www.thetimes.co.uk/tto/business/markets/china/article2614975.ece.
31. Janis Kelly, "Vioxx Hearing Raises Questions about what Merck Knew and When," *Medscape*, November 25, 2004.
32. "An Open Letter from Merck," by Raymond V. Gilmartin, Chairman, President and CEO, http://www.thepinksheet.com/~/media/Images/Publications/Archive/The%20Pink%20Sheet%20Daily/2004/11/12/14041112005/041115_vioxx_openletter.pdf.
33. MSNBC, "FDA Failed Public on Vioxx, Scientist Says," MSNBC, November 19, 2004, http://www.msnbc.msn.com/id/6520630/ns/health-arthritis/t/fda-failed-public-vioxx-scientist-says/.
34. Ralph Nader, *Unsafe at Any Speed: The Designed-In Dangers of the American Automobile* (New York: Grossman Publisher, 1965).
35. *Wall Street Journal*, "Vioxx Testimony," November 18, 2004, http://online.wsj.com/article/SB110079497092278082.html.
36. Ibid.
37. "FDA Drug Screening Measures under Intense Scrutiny," PBS Online NewsHour, Transcript, November 23, 2004, http://www.pbs.org/newshour/bb/government_programs/july-dec04/fda_11-23.html.
38. Alex Berenson, Gardiner Harris, Barry Meier, and Andrew Pollack, "Despite Warnings, Drug Giant Took Long Path to Vioxx Recall," *New York Times*, November 14, 2004, http://

www.nytimes.com/2004/11/14/business/14merck.html?_r=1&ei=5065&en=498edd99049ef255&ex=1101013200&partner=MYWAY&pagewanted=print&.

39. U.S. Department of Justice, "U.S. Pharmaceutical Company Merck Sharp & Dohme to Pay Nearly One Billion Dollars over Promotion of Vioxx®," Office of Public Affairs Press Release, November 22, 2011, http://www.justice.gov/opa/pr/2011/November/11-civ-1524.html.

40. R. V. Gilmartin, "For 100 Years, Patients First," Merck, 2004.

41. Raymond V. Gilmartin, "CEOs Need a New Set of Beliefs," HBR Blog Network/HBS Faculty, *Harvard Business Review*, September 26, 2011, http://blogs.hbr.org/hbsfaculty/2011/09/ceos-need-a-new-set-of-beliefs.html.

42. *Economist*, "The LIBOR Scandal: The Rotten Heart of Finance," July 7, 2012, http://www.economist.com/node/21558281?fsrc=scn/tw_ec/the_rotten_heart_of_finance.

43. Nelson D. Schwartz and Jessica Silver-Greenberg, "JPMorgan's Trading Loss Is Said to Rise at Least 50%," *New York Times*, May 16, 2012.

44. Ben W. Heineman Jr., "Only Government Intervention Can Stop Corrupt Capitalism," *The Atlantic*, July 11, 2012, http://www.theatlantic.com/politics/archive/2012/07/only-government-intervention-can-stop-corrupt-capitalism/259687/#bio; Ben W. Heineman Jr., *High Performance with High Integrity (Memo to the CEO)* (Cambridge, MSA: Harvard Business School Press, 2008).

45. Ibid.

46. Ibid.

47. Barclays, "Citizenship at Barclays," http://group.barclays.com/about-barclays/citizenship.

48. GlaxoSmithKline, "Responsibility," http://www.gsk.com/responsibility.html.

49. CED, "Restoring Trust in Corporate Governance: The Six Essential Tasks of Boards of Directors and Business Leaders," CED Report, January 2010, http://www.law.harvard.edu/programs/plp/pdf/Restoring_Trust_in_Corporate_Governance.pdf.

50. David B. Farber, "Restoring Trust After Fraud: Does Corporate Governance Matter?" http://leeds-faculty.colorado.edu/bhagat/RestoringTrustAfterFraud.pdf. Also see Elisse B. Walter, "Restoring Investor Trust through Corporate Governance," Remarks Before the Practicing Law Institute by SEC Commissioner, http://www.sec.gov/news/speech/2009/spch021809ebw.htm.

51. Ernest & Young, "12th Global Fraud Survey: Growing Beyond: A Place for Integrity," http://www.ey.com/GL/en/Services/Assurance/Fraud-Investigation---Dispute-Services/Global-Fraud-Survey---a-place-for-integrity.

52. Mitchell Ogisi, "Majority Worldwide Sees Widespread Corruption in Businesses," Gallup, May 10, 2012, http://www.gallup.com/poll/154571/Majority-Worldwide-Sees-Widespread-Corruption-Businesses.aspx.

53. *The Telegraph*, "LIBOR Scandal: How I Manipulated the Bank Borrowing Rate," July 1, 2012, http://www.telegraph.co.uk/finance/newsbysector/banksandfinance/9368430/Libor-scandal-How-I-manipulated-the-bank-borrowing-rate.html.

6. CHINA'S AUTHORITARIANISM CHALLENGES AMERICAN DEMOCRATIC EXCEPTIONALISM

1. Katinka Barysch, Charles Grant, and Mark Leonard, "Embracing the Dragon: Can the EU and China Be Friends?" Center for Economic Studies, http://www.cesifogroup.de/portal/pls/portal/docs/1/1206773.PDF.

2. Jerry Oppenheimer, *Toy Monster: The Big, Bad World of Mattel* (Hoboken NJ: John Wiley & Sons, 2009), 219.

3. Louise Story, "Mattel Official Delivers an Apology in China," *New York Times*, September 22, 2007, http://www.nytimes.com/2007/09/22/business/worldbusiness/22toys.html?pagewanted=print .

4. Tom Fordyce, "The Track Record on Olympic Corruption," BBC, July 30, 2004, http://news.bbc.co.uk/sport2/hi/other_sports/olympics_2012/3939219.stm.

5. Jacques deLisle, "After the Gold Rush: The Beijing Olympics and China's Evolving International Roles," *Orbis*, Foreign Policy Research Institute, Spring 2009, http://www.fpri.org/orbis/5302/delisle.aftergoldrush.pdf.

6. Slavoj Zizek, "How China Got Religion," *New York Times*, October 11, 2007, http://www.nytimes.com/2007/10/11/opinion/11zizek.html?pagewanted=print.

7. Melinda Liu and Isaac Stone Fish, "China's Rulers Party On," *The Daily Beast*, June 30, 2011, http://www.thedailybeast.com/articles/2011/06/30/china-s-human-rights-abuses-continue-despite-ai-weiwei-s-release.html; *New York Times*, "Tibetan Riots Spread Outside Region," March 16, 2008, http://www.nytimes.com/2008/03/16/world/asia/16iht-tibet.4.11148124.html.

8. N. D. Batra, "It's a Dog's Life without Freedom," Global Diplomat Blog, *The Statesman*, October 25, 2005, http://globaldiplomat.blogspot.com/2005/10/who-needs-freedom-if-china-can-do.htm; "Beyond Beijing Olympics," August 5, 2008, http://globaldiplomat.blogspot.com/2008/08/china-today.html.

9. Nick Carey and James B. Kelleher, "Special Report: Does Corporate America Kowtow to China?" *Reuters*, April 27, 2011, http://www.reuters.com/article/2011/04/27/us-special-report-china-idUSTRE73Q10X20110427. Also see Lulu Yilun Chen, "Apple Ordered to Remove Obscene Content from China Store," April 17, 2013, http://www.bloomberg.com/news/2013-04-17/apple-ordered-to-remove-obscene-content-from-china-store.html; Rebecca Greenfield, "Apple's Apology May Have Worked Too Well in China," Atlantic Wire, April 2, 2013, http://www.theatlanticwire.com/technology/2013/04/apple-apology-china/63776/.

10. Pew Global, "Growing Concerns in China about Inequality, Corruption," Pew Research Center, October 10, 2012, http://www.pewglobal.org/2012/10/16/growing-concerns-in-china-about-inequality-corruption.

11. Howard W. French, "China Allows More Protests in Shanghai against Japan," *New York Times*, April 17, 2005, http://www.nytimes.com/2005/04/17/international/asia/17china.html.

12. Yinan He, "History, Chinese Nationalism and the Emerging Sino-Japanese Conflict," *Journal of Contemporary China* 16, no. 50 (February 2007): 1–24, http://www.stephen-t.com/pdf/Y%20He%20-%20Journal%20of%20Contemporary%20China,%202007%20-%20princeton.edu.pdf.

13. James Fallows, "'Be Nice to the Countries That Lend You Money,'" *The Atlantic*, December 2008, http://www.theatlantic.com/magazine/archive/2008/12/-8220-be-nice-to-the-countries-that-lend-you-money-8221/7148/.

14. Ellen Nakashima, "U.S. Publicly Calls on China to Stop Commercial Cyber-Espionage, Theft of Trade Secrets," *Washington Post*, March 11, 2013, http://articles.washingtonpost.com/2013-03-11/world/37620683_1_mandiant-china-source-code.

15. Mark Helprin, "Beyond the Rim," *Wall Street Journal*, December 13, 2004, http://online.wsj.com/article/0,,SB110290142784998081,00.html.

16. *Space Daily*, "Tight Security at Launch Site as China Counts Down to Lift-Off," Jiuquan, China (AFP), October 14, 2003, http://www.spacedaily.com/2003/031014112047.karyh00x.html; Rob Gifford, *China Road: A Journey into the Future of a Rising Power* (New York: Random House, 2007); N. D. Batra, "China Rising," Global Diplomats Blog, *The Statesman*, October 22, 2003, http://globaldiplomat.blogspot.com/2009/02/cyber-age.html.

17. Joseph Kahn, "Reaching for Moon, China Works to Put Astronauts in Orbit," *New York Times*, March 14, 2003, http://www.nytimes.com/2003/03/14/world/reaching-for-moon-china-works-to-put-astronauts-in-orbit.html?pagewanted=print&src=pm.

18. Xubei Luo and Nong Zhu, "Rising Income Inequality in China: A Race to the Top," The World Bank, August 25, 2008, http://econ.worldbank.org/external/default/main?pagePK=64165259&theSitePK=469372&piPK=64165421&menuPK=64166093&entityID=000158349_20080825160141.

19. Dexter Roberts, "China's Growing Income Gap," *Bloomberg BusinessWeek*, January 27, 2011, http://www.businessweek.com/magazine/content/11_06/b4214013648109.htm.

20. *USA Today*, "China Gives More Press Freedom—For Food Safety," *Associated Press*, May 15, 2011, http://yourlife.usatoday.com/fitness-food/safety/story/2011/05/China-gives-press-more-freedom-for-food-safety---/47229504/1.

21. Zhiwu Chen, "Free Press Could Help China's Economy," *Financial Times*, September 19, 2005, http://www.ft.com/cms/s/1/97db5d0e-293f-11da-8a5e-00000e2511c8.html.

22. Ibid.

23. N. D. Batra, "Can China Crush Tibet?," Global Diplomat Blog, *The Statesman*, March 25, 2008, http://globaldiplomat.blogspot.com/2008/03/can-china-crush-tibet.html; Also, "All Eyes on Tibet," *AsiaMedia*, UCLA International Institute, March 25, 2008, http://www.international.ucla.edu/article.asp?parentid=89487.

24. Malcolm Byrne, ed., *The 1956 Hungarian Revolution: A History in Documents*, A National Security Archive Electronic Briefing Book, November 4, 2002, http://www.gwu.edu/~nsarchiv/NSAEBB/NSAEBB76/.

25. Finlo Rohrer, "Why Does the West Love the Dalai Lama?" *BBC News Magazine*, February 18, 2010, http://news.bbc.co.uk/2/hi/8521957.stm.

26. *New York Times*, "Bush Silent, but Others Speak Out on Tibet Crackdown," March 22, 2008, http://www.nytimes.com/2008/03/22/world/asia/22prexy.html.

27. *Economist*, "Nothing to Celebrate," February 15, 2013, http://www.economist.com/blogs/analects/2013/02/tibetans.

28. *Der Spiegel*, "German Minister Defends Meeting with Dalai Lama," http://www.phayul.com/news/article.aspx?id=21270&t=1.

29. Ibid.

30. Andrew Pierce and Richard Spencer, "Prince of Wales to Meet Dalai Lama in London," *The Telegraph*, March 20, 2008, http://www.telegraph.co.uk/news/worldnews/1582329/Prince-of-Wales-to-meet-Dalai-Lama-in-London.html.

31. I discussed this issue in several newspaper articles and blogs. See N. D. Batra, "China's Enemy in Cyberspace," Global Diplomat Blog, *The Statesman*, April 1, 2008, http://globaldiplomat.blogspot.com/2008/.../chinas-enemy-not-in-dharamsa.

32. N. D. Batra, "Disconcerting Telepresence," *Times of India*, January 25, 2012, 12.00AM IST rticles.timesofindia.indiatimes.com/2012-01-25/edit-page/30659754_1_human-activity-american-jobs-digital.

33. Ibid.

34. H. A. Innis, "The Bias of Communication," *The Canadian Journal of Economics and Political Science / Revue canadienne d' Economique et de Science politique* 15, no. 4 (November 1949): 457–76, http://www.jstor.org/pss/138041.

35. UNPO, "Tibet: Demographic Aggression Continues," http://www.unpo.org/article/7947.

36. Amnesty International, "China: Authorities Must Drop Charges against Human Rights Activist," December 22, 2011, http://www.amnestyusa.org/news/news-item/china-authorities-must-drop-charges-against-human-rights-activist.

37. Committee to Protect Journalists, "Attacks on the Press in 2011," http://cpj.org/2012/02/attacks-on-the-press-in-2011.php.

38. N.D. Batra, "Unbowed and Unafraid," *The Statesman*, January 15, 2010, http://www.thestatesman.net/index.php?id=317575&option=com_content&catid=39.

39. Cryptome, "Eyeballing Tiananmen Massacre," http://cryptome.org/cn/tk/tiananmen-kill.htm.

40. Charter 08, http://www.charter08.eu/2.html.

41. Michael Wines and Sharon LaFraniere, "During Visit, Obama Skirts Chinese Political Sensitivities," *New York Times*, November 17, 2009, http://www.nytimes.com/2009/11/18/world/asia/18china.html.

42. *New York Times*, "Mrs. Clinton's Unwavering Words," September 6, 1995, http://www.nytimes.com/1995/09/06/opinion/mrs-clinton-s-unwavering-words.html.

43. Nobelprize.org, "I Have No Enemies: My Final Statement," Nobel Lecture in Absentia, December 10, 2010, Statement of December 23, 2009, Read by Liv Ullmann, Translation By HRIC, http://www.nobelprize.org/nobel_prizes/peace/laureates/2010/xiaobo-lecture.html; N. D. Batra, "Unbowed and Unafraid," *The Statesman*, January 18, 2010, http://

www.thestatesman.net/index.php?option=com_content&view=article&id=317575:unbowed-and-unafraid&catid=39:perspective&from_page=search.

44. Susan L. Shirk, ed., *Changing Media, Changing China* (New York: Oxford University Press, 2011).

45. Michael Bristow, "China's Internet 'Spin Doctors,'" BBC, December 16, 2008, http://news.bbc.co.uk/2/hi/7783640.stm.

46. Michael Wines, "Crackdown on Chinese Bloggers Who Fight the Censors with Puns," *New York Times*, May 28, 2012, http://www.nytimes.com/2012/05/29/world/asia/china-cracks-down-on-its-cagey-web-critics.html.

47. BBC, "China Condemns Decision by Google to Lift Censorship," March 23, 2010, http://news.bbc.co.uk/2/hi/asia-pacific/8582233.stm.

48. Joseph Stiglitz and Roumeen Islam, "Why a Free Press Boosts Economic Development," *Financial Review* 21.11.02.

49. Shirk, *Changing Media, Changing China*, 4.

50. Ibid.

51. Isabella Bennett, "Media Censorship in China," Council on Foreign Relation, March 7, 2011, http://www.cfr.org/china/media-censorship-china/p11515; Robert McMahon and Isabella Bennett, "U.S. Internet Providers and the 'Great Firewall of China,'" Council on Foreign Relations, February 23, 2011, http://www.cfr.org/china/us-internet-providers-great-firewall-china/p9856.

52. "Jon Huntsman Predicts 'Major Problems' Ahead for China," *Huffington Post*, June 22, 2011, http://www.huffingtonpost.com/2011/06/20/jon-huntsman-predicts-major-problems-for-china_n_880402.html.

53. *Guardian*, "China Unrest as Hundreds Protest against Taizhou Land Grabs," June 16, 2011, http://www.guardian.co.uk/world/2011/jun/16/china-riots-taizhou.

54. *Financial Times*, "China's Restive Migrant Workers," June 19, 2011, http://www.ft.com/intl/cms/s/0/b2239ad8-9a9e-11e0-bab2-00144feab49a.html#axzz1PvNr7lHv.

55. Max Fisher, "Is the 'China Model' Failing?" *The Atlantic*, June 21, 2011, http://www.theatlantic.com/international/archive/2011/06/is-the-china-model-failing/240773/.

56. Steve Lohr, "When Innovation, Too, Is Made in China," *New York Times*, January 1, 2011, http://www.nytimes.com/2011/01/02/business/02unboxed.html?_r=2&nl=technology&emc=techupdateema3.

57. Ibid.

58. PBS, "Who Is Afraid of Ai Weiwei," *PBS Frontline*, http://www.pbs.org/wgbh/pages/frontline/ai-wei-wei/.

59. Xiao Qiang, "The Rise of Online Public Opinion and Its Political Impact," in *Changing Media, Changing China*, ed., Susan L. Shirk (New York: Oxford University Press, 2011), 206; Freedom House, "Freedom on the Net 2011: China," http://www.wbasc.net/wp-content/uploads/2011/07/China2011.pdf.

60. Shirk, *Changing Media, Changing China*, 210.

61. Ibid.

62. Qian Gang, "China's Reporters Push the Boundaries," *Wall Street Journal*, August 3, 2011, http://online.wsj.com/article/SB10001424053111903520204576484241060215576.html?mod=googlenews_wsj.

63. Ibid.

64. Tania Branigan, "Chinese Anger over Alleged Cover-Up of High-Speed Rail Crash," Guardian.co.uk, http://www.guardian.co.uk/world/2011/jul/25/chinese-rail-crash-cover-up-claims.

65. Andrew J. Nathan, "What China Wants: Bargaining with Beijing," *Foreign Affairs* 90, no. 4 (July/August 2011): 156.

7. POLITICAL INNOVATIONS OF THE ASPIRATIONAL SOCIETY

1. *Rebirth*, Project Rebirth, January 2011, http://projectrebirth.org/film/.

2. Wayne Barrett and Dan Collins, *Grand Illusion: The Untold Story of Rudy Giuliani and 9/11* (New York: Harper Collins, 2006).

3. Lawrence Wright, *Looming Tower: Al-Qaeda and the Road to 9/11* (New York: Knopf, 2006).

4. George W. Bush, "9/11 Address to the Nation," American Rhetoric, September 11, 2001, http://www.americanrhetoric.com/speeches/gwbush911addresstothenation.htm.

5. "African Embassy Bombings," Online NewsHour, PBS.org, http://www.pbs.org/newshour/bb/africa/embassy_bombing/.

6. "Uncommon Knowledge: George W. Bush," The Hoover Institution, July 18, 2012, http://www.youtube.com/user/HooverInstitution?feature=watch; George W. Bush, *Decision Points* (New York: Broadway Paperbacks, 2011).

7. George W. Bush, "The President's News Conference," November 7, 2002, http://www.presidency.ucsb.edu/ws/index.php?pid=63868#axzz1RKQXjt2A.

8. "Security Council Holds Iraq in 'Material Breach' of Disarmament Obligations, Offers Final Chance to Comply, Unanimously Adopting Resolution 1441 (2002)," Security Council, 4644th Meeting (AM), August 11, 2002, http://www.un.org/News/Press/docs/2002/SC7564.doc.htm.

9. Elizabeth Bumiller, "The 2002 Election: The President; After G.O.P. Triumph, Bush Sets Security Department and Tax Cut as Priorities," *New York Times*, November 8, 2002, http://www.nytimes.com/2002/11/08/us/2002-election-president-after-gop-triumph-bush-sets-security-department-tax-cut.html?pagewanted=all&src=pm.

10. For example, James Champy, the author of *X-Engineering The Corporation*, was quoted as having said, "I once was a believer in real, radical movements. You can't just blow them up and start over with a clean sheet. Faced with the complexity you have, you must step back, look at the pieces, and determine where are the leverage points and then drive like heck and focus. If you can do that, you can pull along the rest of the enterprise in five to 10 years." See Eric Chabrow, "Homeland Security Department Will Need Time to Come Together," Informationweek.com, November 5, 2002, http://www.informationweek.com/homeland-security-department-will-need-t/6504220.

11. John F. Burns, "Threats and Responses: Baghdad; A Top Iraqi Aide Defies U.S. to Find Proof of Weapons," *New York Times*, December 09, 2002, http://www.nytimes.com/2002/12/09/world/threats-responses-baghdad-top-iraqi-aide-defies-us-find-proof-weapons.html?pagewanted=all&src=pm.

12. "BBC Breakfast with Frost Interview: Jack Straw MP Foreign Secretary," BBC December 8, 2002, http://news.bbc.co.uk/2/hi/programmes/breakfast_with_frost/2555907.stm.

13. John F. Burns, "Threats and Responses."

14. "Iraq Urges US to Let World See Evidence," *Times*, December 9, 2002, http://www.timesonline.co.uk/tol/news/world/article799948.ece.

15. "Colin Powell Demands Answers over Curveball's WMD Lies," *The Guardian*, February 16, 2011, http://www.guardian.co.uk/world/2011/feb/16/colin-powell-cia-curveball.

16. "Inspector: Iraq Had No WMD before Invasion," MSNBC, June 10, 2004, http://www.msnbc.msn.com/id/6190720/ns/world_news-mideast_n_africa/t/inspector-iraq-had-no-wmd-invasion/#.UAmzQmlSSmI.

17. Patrick E. Tyler, "U.N. Chief Ignites Firestorm by Calling Iraq War 'Illegal,'" *New York Times*, September 17, 2004, http://www.nytimes.com/2004/09/17/international/europe/17nations.html.

18. Thomas L. Friedman, "The Battle of the Pump," *New York Times*, October 7, 2004, http://www.nytimes.com/2004/10/07/opinion/07friedman.html.

19. Niall Ferguson, "The Way We Live Now: War Names," *New York Times*, December 15, 2002, http://www.nytimes.com/2002/12/15/magazine/15WWLN.html?pagewanted=all.

20. Christopher Bassford, *Clausewitz in English: The Reception of Clausewitz in Britain and America* (Oxford, UK: Oxford University Press, 1994).

21. Ferguson, "The Way We Live Now."

22. André Beaufre, *The Suez Expedition 1956* (New York: Praeger, 1969).

23. Carl Robichaud, "Failings of the Rumsfeld Doctrine," *Christian Science Monitor*, September 21, 2006, http://www.csmonitor.com/2006/0921/p09s02-coop.html.

24. Ferguson, "The Way We Live Now."
25. Pervez Musharraf, *In the Line of Fire: A Memoir* (New York: Simon and Schuster, 2006); *New York Times*, "Pervez Musharraf," http://topics.nytimes.com/top/reference/timestopics/people/m/pervez_musharraf/index.html.
26. CNN, "Bush: 'We're Slowly But Surely Achieving Our Objective,'" March 23, 2003, http://articles.cnn.com/2003-03-23/us/sprj.irq.bush.transcript_1_oil-fields-coalition-forces-objective?_s=PM:US.
27. NBC, "Interview with Vice President Dick Cheney," *Meet the Press*, Transcript, Sunday, March 16, 2003, https://www.mtholyoke.edu/acad/intrel/bush/cheneymeetthepress.htm.
28. Michael Moore, *Stupid White Men . . . And Other Sorry Excuses for the State of the Nation!* (New York: Harper Collins, 2004); Michael Moore, in his 2003 Academy Awards acceptance speech. Retrieved July 7, 2011, http://politicalhumor.about.com/cs/quotethis/a/moorequotes.htm.
29. Natana DeLong-Bas, *Wahhabi Islam : From Revival and Reform to Global Jihad* (New York: Oxford University Press, 2008(.
30. Joseph Kahn, "Clinton 'History' Doesn't Repeat Itself in China," *New York Times*, September 24, 2003, http://www.nytimes.com/2003/09/24/world/clinton-history-doesn-t-repeat-itself-in-china.html?pagewanted=all&src=pm.
31. Jon Clifton, "50 Million Adults Worldwide Would Migrate to the U.S.," Gallup, April 20, 2012, http://www.gallup.com/poll/153992/150-million-adults-worldwide-migrate.aspx; Ian Bremmer, "Are State-Led Economies Better?" *Reuters*, July 3, 2012, http://blogs.reuters.com/ian-bremmer/2012/07/03/are-state-led-economies-better/; Frank Langfitt, "China's Rich Consider Leaving Growing Nation," NPR, January 17, 2012, http://www.npr.org/2012/01/17/145314824/chinas-rich-consider-leaving-growing-nation.
32. "Most Chinese Millionaires Want to Emigrate," *China Times*, November 1, 2011, http://www.thechinatimes.com/online/2011/11/1810.html.
33. Mao Tse-Tung, "I believe the United States is nothing but a paper tiger. . . . It seems that the countries of the Americas, Asia and Africa will have to go on quarrelling with the United States till the very end, till the paper tiger is destroyed by the wind and the rain" in "U.S. Imperialism is a Paper Tiger," *Selected Works of Mao Tse-tung*, July 14, 1956, http://www.marxists.org/reference/archive/mao/selected-works/volume-5/mswv5_52.htm.
34. Joseph S. Nye Jr., "The richest country in the world could afford both better education at home and the international influence that comes from an effective aid and information program abroad. What is needed is increased investment in 'soft power,' the complex machinery of interdependence, rather than in 'hard power'—that is, expensive new weapons systems," in "Misleading Metaphor of Decline," *Wall Street Journal*, February 14, 2011, http://online.wsj.com/article/SB10001424052748704358704576118673650278558.html; Joseph S. Nye Jr., "Soft Power: Propaganda Isn't the Way," *New York Times*, January 10, 2003, http://www.nytimes.com/2003/01/10/opinion/10iht-ednye_ed3_.html.
35. Suzanne Nossel, "Smart Power," *Foreign Affairs*, March–April 2004, http://www.foreignaffairs.com/articles/59716/suzanne-nossel/smart-power.
36. Thurston Clarke, *Ask Not: The Inauguration of John F. Kennedy and the Speech That Changed America* (New York: Henry Holt and Co., 2004).
37. A. S. Grove, *Swimming Across: A Memoir* (New York: Hachette Book Group, 2001).
38. Colin Powell, *My American Journey* (New York: The Random House Ballantine Books, 1995), 591–2.
39. Bill Cosby, *Fatherhood* (New York: Doubleday, 1986); Bill Cosby and Alvin F. Poussaint, *Come on, People: On the Path from Victims to Victors* (Nashville, TN: Thomas Nelson, 2007); Lisa Campbell, *Michael Jackson: The King of Pop* (Wellesley, MA: Branden Books, 1993); NBA Encyclopedia, "Michael Jordon," http://web.archive.org/web/20060707083040/http://www.nba.com/history/players/jordan_bio.html.
40. Collin Powell, *My American Journey.*
41. Ibid.
42. Stephanie Chen, "Does Your Name Shape Your Destiny?" CNN, May 26, 2010, http://edition.cnn.com/2010/LIVING/05/26/naming.names.importance/?hpt=Sbin.

43. Tom Baldwin, "The Day the White House Opens Its Doors to Its First Black Incumbent," *Times*, November 10, 2008, http://www.thetimes.co.uk/tto/news/world/americas/article1890435.ece; N. D. Batra, "The Flight of Black Swan," Global Diplomat Blog, *The Statesman*, November 1, 2008, globaldiplomat.blogspot.com/2008_11_01_archive.html.

44. PBS, "The Rise and Fall of Jim Crow," October 1–22, 2002, http://www.pbs.org/wnet/jimcrow/.

45. Jeff Zeleny, "Obama, in His New Role as President-Elect, Calls for Stimulus Package," November 8, 2008, http://www.nytimes.com/2008/11/08/us/politics/08obama.html?pagewanted=print .

46. Jeffrey Goldberg, "The Point of No Return," *The Atlantic*, September 2010, http://www.theatlantic.com/magazine/archive/2010/09/the-point-of-no-return/8186/?single_page=true.

47. I discussed the rise of Barack Obama in a number of articles I wrote for the *Statesman*. See, for example, N. D. Batra, "The American Fantasy: Mr. Obama and the Dream of Exceptionalism," *The Statesman*, February 8, 2010, http://www.thestatesman.net/index.php?option=com_content&view=article&id=319402:special .

48. Bertolt Brecht, *Life of Galileo*, trans. Charles Laughton (New York: Arcade Publishing, 1994).

49. President George W. Bush, State of the Union Address, January 29, 2002, http://georgewbush-whitehouse.archives.gov/news/releases/2002/01/print/20020129-11.html; President Obama's speech at the Nobel Peace Prize ceremony in Oslo, December 11, 2009, http://www.nytimes.com/2009/12/11/world/europe/11prexy.text.html.

50. Donald E. Pease, *The New American Exceptionalism* (Minneapolis, MN: University of Minnesota Press, 2009).

51. Ibid., 155.

52. "The shock and awe" doctrine of total dominance in the battlefield by all the available means to paralyze the enemy's will to fight is associated in the popular imagination with the 2003 invasion of Iraq, though its main characteristics and features were articulated earlier; Harlan K. Ullman and James P. Wade, "Shock and Awe: Achieving Rapid Dominance," National Defense University, 1996, http://www.dodccrp.org/files/Ullman_Shock.pdf.

53. Pease, *The New American Exceptionalism*, 154.

54. Ibid., 155.

55. Ibid., 190.

56. Ibid., 192.

57. Cindy Sheehan, *Dear President Bush* (San Francisco, CA: City Lights Publishers, 2006); Pease, *The New American Exceptionalism*, 201–2.

58. In Sophocles' play *Antigone* (441 BC), the protagonist, Antigone, after Thebes's civil war, buries her rebellious brother Polyneices against the new ruler Creon's edict that the rebel brother's body, instead of being consecrated by holy rites, will remain unburied on the battlefield for worms and vultures to feed on.

59. Robert Graves, *The Greek Myths* (London, UK: Penguin, 1955).

CONCLUSION: FREE SPEECH AS ENERGY

1. Ian Morris, *Why the West Rules—for Now: The Patterns of History, and What They Reveal About the Future* (New York: Farrar, Straus and Giroux, 2010). Morris, "Here Comes the East," *New York Times*, December 21, 2010, http://www.nytimes.com/2010/12/22/opinion/22iht-edmorris22.html?_r=0.

2. The Royal Society, http://royalsociety.org/about-us/history/.

3. *60 Minutes*, "Paralyzed Woman Uses Mind to Move Robotic Arm," http://www.cbsnews.com/8301-18560_16-57559331/paralyzed-woman-uses-mind-to-move-robotic-arm/.

4. Stuart Kauffman, *Investigations* (New York: Oxford University Press, 2000), 48.

5. Eric X. Li, "The Life of the Party: The Post-Democratic Future Begins in China," *Foreign Affairs* 92, no. 1 (2013); Yasheng Huang, "Democratize or Die: Why China's Communists Face Reform or Revolution," *Foreign Affairs* 92, no. 1 (2013).

6. Jonathan Manthorpe, "Illicit Billions Flood out of China," *Edmonton Journal*, October 31, 2012, http://www2.canada.com/edmontonjournal/news/ideas/story.html?id=4ea15f94-ea19-42f8-ad2b-819404c26efc&k=88372'; Pan Kwan Yuk, "China's Rich Head for the Exit," *FT Blog*, May 24, 2012, http://blogs.ft.com/beyond-brics/2012/05/24/chinas-rich-head-for-the-exit/#axzz2ISwQZx7k.

7. Martin Wolf, *Why Globalization Works* (New Haven, CT: Yale University Press, 2004), 25.

8. Stuart Kauffman, *Investigations* (New York: Oxford University Press, 2000), 22.

9. Frank Partnoy and Jesse Eisinger, "What's Inside America's Banks?" *The Atlantic*, January/February 2013, http://www.theatlantic.com/magazine/archive/2013/01/whats-inside-americas-banks/309196/?single_page=true.

10. *Economist*, "Year of the Lawyer," January 5, 2013, http://www.economist.com/news/finance-and-economics/21569053-banks-face-another-punishing-year-fines-and-lawsuits-year-lawyer.

11. Gretchen Morgenson, "How to Cut Megabanks Down to Size," *New York Times*, January 19, 2013, http://www.nytimes.com/2013/01/20/business/a-fed-voice-asking-to-cut-megabanks-down-to-size.html?emc=eta1&_r=0.

12. For a discussion on dependent origination or interdependent coarising, see Joanna Macy, *Mutual Causality in Buddhism and General Systems Theory: The Dharma of Natural Systems* (Albany, NY: State University of New York Press, 1991).

13. Percy Bysshe Shelley, "Prometheus Unbound," http://www.calstatela.edu/faculty/jgarret/560/shelley-pu.pdf.

> To suffer woes which Hope thinks infinite;
> To forgive wrongs darker than death or night;
> To defy Power, which seems omnipotent;
> To love, and bear; to hope till Hope creates
> From its own wreck the thing it contemplates;
> Neither to change, nor falter, nor repent;
> This, like thy glory, Titan, is to be
> Good, great and joyous, beautiful and free;
> This is alone Life, Joy, Empire, and Victory.

Index

About the Author

Narain D. Batra, PhD, is professor of communications at Norwich University, Vermont, where he teaches media law, ethics, criticism, and social media and the Internet. He also teaches global corporate diplomacy in the graduate program in the Master of Arts in Diplomacy at Norwich University.

Author of several books, Professor Batra's most recent book is *Digital Freedom: How Much Can You Handle?* (Rowman & Littlefield Publishers, Inc., 2008).

Professor Batra wrote a weekly column for *The Statesman* (India, 1995–2010). The weekly column "Cyber Age" was distributed to several Asian newspapers through Asia News Network, Bangkok, Thailand. He published more than seven hundred columns on American culture, communications technology, mass media, diplomacy, free speech, business, South Asia, and China. Now Professor Batra writes for *The Times of India* and other newspapers.

Professor Batra was editor and publisher of a monthly magazine *People's Times* (India, 1970–1977). A prolific writer, he publishes a blog, *Global Diplomat*, at http://globaldiplomat.blogspot.com, and he tweets regularly.

Professor Batra is a public lecturer and an avid golfer.

www.ingramcontent.com/pod-product-compliance
Lightning Source LLC
LaVergne TN
LVHW101446200926
848483LV00001B/64

* 9 7 8 1 4 4 2 2 2 5 8 7 9 *